LEARNING FROM DISASTERS: A MANAGEMENT APPROACH

Brian Toft
Simon Reynolds

Perpetuity Press

First published in 1994 by Butterworth-Heinemann Ltd

This second edition published in 1997 by Perpetuity Press Limited
PO Box 376, Leicester, LE2 3ZZ

British Library Cataloguing in Publication Data.
A catalogue record for this book is available from the British Library

ISBN 1-899287-05-1

Perpetuity Press

Printed and bound in Great Britain by Biddles Ltd, Guildford and King's Lynn

Contents

Foreword

Foreword to the Second Edition of "Learning from Disasters"

In the latter half of the twentieth century giant strides have been made in putting into place the mechanics for improving safety. Legislatory and regulatory frameworks which have been adopted by most industrialised nations have prevented some of the worst excesses of the past. Responsible organisations no longer have a cavalier attitude to safety. And yet disasters and accidents of all kinds continue to occur and the scope for doing better – to reduce the still enormous waste of resources and the accompanying human misery – remains very large indeed.

So why does the body corporate find it so hard to learn from past mistakes? This book provides compelling evidence that organisations will only be able to absorb the lessons of history if they understand the full range of cultural forces at work as well as the technicalities of any accident. Much of the focus of normal investigation is on the operations which immediately precede a disaster There is rarely an understanding that operators are the inheritors of system defects, incorrect installation, faulty designs, wrong materials and above all bad management. Attention tends to concentrate on the final garnish rather than the basic ingredients of a lethal brew which has already been long in the cooking.

An inability to learn from hindsight should be a matter of concern to every Board of Directors. The analysis contained in this volume suggests that it is only by looking at disasters as an overall system failure that the true lessons will emerge. By helping us to overcome the inhibitions to learning from past experience it opens up the avenues to better risk management and avoidance of major accidents. Recognising latent errors which lie dormant in the 'system' together with the insidious and often unforeseeable ways in which they combine to breach defences at some critical moment is a key issue.

Understanding the complex socio-technical patterns involved in accidents of all kinds will be greatly enhanced by this excellent book. I commend it to everybody who wants to gain an insight into the forces at work which lead up to disasters.

Sir Neville Purvis KCB
Director General
British Safety Council

Preface

The original edition of this book demonstrated how organisations can learn from disasters, and this is still the fundamental message of the current work. However, given that the management of risk within organisations has become a topic of increasing interest to the business community the authors have republished this book, with the addition of a new opening chapter that provides a risk management framework from within which the original work can now be viewed.

In the intervening period since the publication of our original book a second edition of Barry Turner's seminal work, *Man-Made Disasters*, has been published by Butterworth-Heinemann, with additional material by Nick Pidgeon. To those interested in understanding how catastrophic failures occur in socio-technical systems we strongly commend this volume.

It may interest readers to know that Barry Turner was Brian Toft's PhD supervisor at Exeter University where he was a significant influence on Brian's academic and personal development. This is perhaps reflected in the fact that Barry's doctoral thesis was entitled 'The Failure of Foresight', while Brian's was entitled 'The Failure of Hindsight'. Sadly Barry Turner died on 24 February 1995 at his home in Chiswick. He will be greatly missed.

Brian Toft
Simon Reynolds

About the authors

Professor Brian Toft BA(Hons) Dip Comp Sci(Cantab) PhD MIOSH FIIRSM FICD FIRM FRSA

Professor Brian Toft is currently a Senior Technical Consultant and Assistant Director at Sedgwick UK Risk Services Limited, a Visiting Professor in Risk Management at Bolton Business School, External Examiner for the MSc degree course in Risk Management and Safety Technology at Aston University, External Examiner for the MSc degree course in Risk, Crisis and Disaster Management at Leicester University, an Honorary Visiting Fellow in Crisis Management at Bradford University, Member of the Board of Governors of the International Institute of Risk and Safety Managers (IIRSM), the Institute of Civil Defence and Disaster studies representative on the National Steering Committee for Warning & Informing the Public During Emergencies, a member of the Institution of Occupational Safety and Health's Policy Development Committee, and the Institute of Risk Management's representative on the British Standard Institute Committee dealing with Occupational Health and Safety.

Brian has published many papers on the subject of risk management and is a member of the Editorial Board of the international journal *Risk, Decision and Policy*.

Simon Reynolds BSc

Simon Reynolds took a mechanical engineering degree, and worked for several years in the power engineering industry. He turned to journalism, gaining experience as an editor with an on-line subscription service providing daily political and macro-economic news. He has since specialised in writing about insurance, banking and risk management issues. For the last seven years he has worked as a freelance, and he regularly contributes to a number of magazines and newsletters.

Chapter 1

The management of risk

Introduction

In this chapter theoretical models and practical arguments will be developed to show that risk management is a cost-effective way of protecting organisations from both human and financial losses. The development of risk management as a discipline is discussed including some of the problems associated with its practical application within organisations.

1.1 Development of risk management

The activity of risk management is the embodiment of the old adage that 'an ounce of prevention is better than a pound of cure'. That is, management should be proactive and attempt to identify and measure hazards and control the risk of their occurrence before they occur.

Unfortunately, there are still many managers in organisations who appear to hold the belief that risk management is some kind of secret esoteric art form whose credibility is just slightly better than that of black magic: that it is an activity outside 'normal' management practices. We argue that such a view is based on myth rather than fact, and that once an organisation has recognised this then it is, at least theoretically, in a position to improve its performance by either preventing or reducing the potential for losses to which it is exposed.

While the concept of risk taking can be traced back to the early Greek and Arab civilisations, the idea of attempting to manage organisational risks is a relatively new concept. A few far-sighted organisations were implementing some of the techniques that now come under the banner of risk management before the turn of the 20th century, particularly in the field of safety management and employee welfare. However, the notion of risk management as a comprehensive, proactive and mainstream management task only gained currency during the middle 1950s and early 1960s, driven by organisations based in the United States of America[1].

An indicator of the maturity of any profession is the extent of its educational provision. While there are some academic and professional courses in the subject of risk management, such as those offered by the Institute of Risk Management in the United Kingdom, there is not the plethora of courses that exists for general management studies such as Master in Business Administration degrees. Thus, the formal and explicit management of fortuitous organisational risks is not a mature activity and, as a consequence, it does not possess the same kind of legitimacy as is accorded to such concepts as total quality management or recognised international quality systems standards such as the ISO 9000 series.

In part this professional immaturity may be attributed to the complexity of the topic. Risks are not concrete entities like computers or motor vehicles, which can be studied largely without subjective bias. Risks cannot be measured in objective, unambiguous terms, for any assessment of them is based on perceptions that are neither neutral nor value free. Otway and Pahner observe that, in individuals:

The perception of risks is a crucial factor in forming attitudes; obviously people respond to a threatening situation based upon what they perceive it to be.[2]

1

Douglas and Wildavsky[3] argue in a similar vein that different societies, and the individuals of which they are composed, create their own sets of criteria against which the risks associated with a particular hazardous circumstance will be interpreted and 'measured'. The use of such social and individual reference schemata suggests that the risks perceived by a given society or individual are not objective but subjective.

This theme has subsequently been elaborated on by several authors. For example, Reid likewise proposes that: '...it is unrealistic to presume that the fundamental processes of risk assessment are objective'.[4] While Shrader-Frechette[5] has stated that: 'All judgements about hazards or risks are value-laden'. The most radical position has been taken by Paul Slovic[6] who has suggested that: 'There is no such thing as "real risk" or "objective risk".' Drawing upon these assertions, it can be postulated that to some extent, all risks are subjective in nature, as suggested by Pidgeon et al.[7]

Indeed, risk is like the Cheshire Cat in that it can suddenly materialise. For example, when he was Chancellor of the Exchequer, Norman Lamont stated that the UK would not leave the European exchange rate mechanism (ERM). He was forced to recant shortly afterwards when there was a run on sterling caused by speculators, and the UK subsequently left the ERM. Either he did not perceive the risks associated with speculators, or if he did, he must have assumed that they were manageable.

Similarly, Barings, the merchant bank, suffered massive financial losses when its directors failed to recognise the risks associated with flawed managerial controls in its Singapore office. The *Financial Times*[8] reported that internal auditors had warned the directors of the threat in August 1994. The board appears to have believed that the risk was so small as to make remedial measures inappropriate, although hindsight shows otherwise.

Other evidence to support this view is to be found in an empirical study into the perception of occupational risks by three different groups of British mineworkers. The researchers, Rushworth et al. Cited by Symes,[9] came to the conclusion that the principal reasons for the workers engaging in unsafe behaviour were:

- they were unaware of the hazard – an activity was not known to be risky;

- they underestimated the risk of injury associated with a particular hazard; and

- expediency, urgency or convenience (that is, although risky, the activity was felt to be worth the risk).

Seccombe and Ball,[10] in a study into the frequency and severity of back injuries among nurses, found that of those who had sustained back injuries at work 41% 'felt that it was always a rush to get work done'.

1.2 Good management is risk management

The difficulties caused by the subjective nature of risk management do not mean that it is any more mysterious an activity than any other form of management. The management of any organisation, in its simplest form, can be envisaged as being three separate, but interrelated components: business management, logistical support and operational management as in Figure 1.1.

Business management is concerned, for example, with getting the orders for the product or services to be provided by a company or with ensuring customer invoices are sent out.

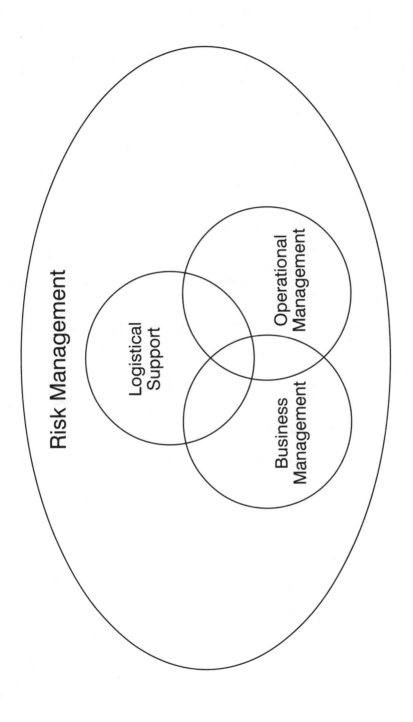

Figure 1.1 Corporate Risk Management

Operational management is about making sure the product or service is produced and carried out to the required specification. Logistical support management tries to ensure that those elements required to support both business and operational activities – for example, people, computers, faxes, telephones etc. – are present.

Each component is designed to ensure that all the organisation's targets are achieved and that unwanted events do not occur. Thus each component should be able to remove, control or reduce the risk of an unwanted event, or ameliorate the consequences if one occurs.

Therefore, it can be postulated that risk management is not an activity which is separate from those which take place in 'mainstream' management but the raison d'être for all management.

1.3 Problems in risk management

There are a number of problems associated with risk management:

- Many of the hazards that plague organisations are frequently geographically dispersed, typically each individual organisation afflicted by a particular hazard believing that they are the only ones to suffer from it. Therefore, it often takes time before the 'real' extent of a particular hazard comes to light.

- The threat of litigation often prevents organisations from revealing or sharing information on actual and potential hazards.

- An organisation's staff are often embarrassed, or afraid of having their employment terminated, if they reveal that they were responsible (or they feel responsible) for a potential or realised hazard. Consequently staff tend not to inform management of any errors which they might have made. This reluctance to advertise any human factor errors unfortunately also includes errors that led to 'near miss' incidents.

- Much of the hazard and risk information that is currently available is not in a form immediately ready for risk pattern or trend analysis.[11]

- There are difficulties in disseminating the lessons learned from unwanted organisational events.

- There is a tendency for the management of some organisations not to take advantage of the lessons others have learned through experience.[12]

- It is impossible for anyone to specify what they do not know.

- While the past is 'fixed' in time there are multiple future realities and thus organisations cannot rely upon past achievements to unerringly predict future performance.

Few theoretical models have been developed to aid our understanding of organisational failures and of methods that might promote organisational success or hazard prevention.

1.4 Prevention of failure model

An attempt to address the problem of the lack of a theoretical model is shown below in Figure 1.2. This simple conceptual model tries to capture some of the fundamental system elements that appear to be found in the generation of organisational failures and, secondly, illustrates two different kinds of actions that appear to affect organisational performance.

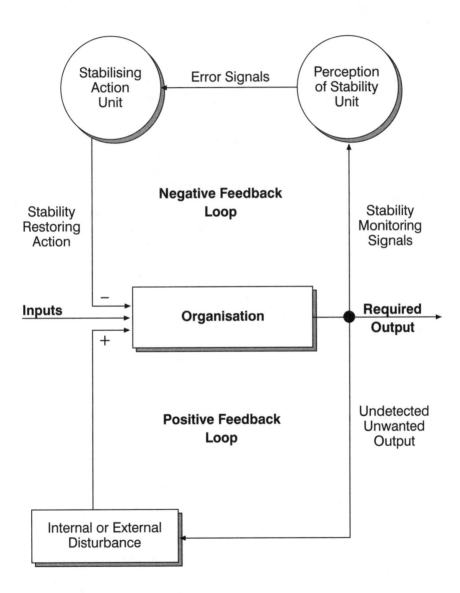

Figure 1.2 Prevention of Failure Model

In the model the Perception of Stability Unit (PSU) and the Stabilising Action Unit (SAU) are concepts and thus their functions can be envisaged as being undertaken by a human being, by a machine or by a combination of both.

Generally, an organisation will have a set of parameters that the output of its activities must attain for the finished 'product' or 'service' to be considered satisfactory. The comparison between the required parameter values and those parameter values of a particular piece of organisational output is typically maintained by a PSU. This unit compares the output of the 'production' process against some desired standards.

Providing that the product or service meets the standard then that product or service is considered acceptable. However, if the product or service should fail to meet the required standard the PSU is designed to send an error signal to the SAU. Upon receiving an error signal the SAU will analyse the signal and then select from its repertoire of allowed behaviours the most appropriate course of action so that the error affecting the product or service is eliminated or reduced to within tolerance limits.

For example, suppose the required output of an organisation is a product that has a mass of one kilogram. If an internal or external disturbance should cause the mass of the items to increase, then the PSU would detect the problem, transmit an error signal to the SAU, which would then reduce the raw material being input to the process by exactly the excess amount. This type of control action is called Negative Feedback control since the action of the system in response to an error signal is always in the opposite sense to the error causing the out of specification condition.

On the other hand, if a disturbance was not recognised by an organisation's PSU then no remedial control action could be taken and the unwanted error condition would increase in size. This form of system behaviour is called Positive Feedback. Consider, for example, how an uncontrolled fire grows in both size and intensity until a flashover occurs when sufficient oxygen and combustible material are present.

While this model is elementary it does provide a number of insights relating to the prevention of failures. Firstly, the model highlights the need for management to continually search for areas within the organisation where negative feedback control systems might be utilised to ensure that positive feedback cycles do not escalate minor errors into full-scale disasters. Secondly, the model suggests the need for management to ensure that the negative feedback systems that they already have in place (for example, supervisors monitoring the work performance of staff) are working as intended.

Thirdly, the PSU needs to be able to recognise the widest possible range of 'out of specification' conditions, as any failure by the Unit to perceive an error and its significance could ultimately lead to the system's destruction. The danger associated with an organisation failing to ensure that negative feedback control action does take place appropriately is clearly spelled out by the findings of the investigation into the multiple train collision at Clapham Junction, on 12 December 1988, in which 35 passengers died and approximately 500 were injured. It was stated that:

> An independent wire count could and should have prevented the accident. The responsibility for the accident does not for a moment lie, as Mr Hemingway seemed to believe it did, solely upon his shoulders. His were the original errors, but they should have been discovered and neutralised by the processes of supervision and testing.[13]

Unfortunately the 'original errors' were not discovered and neutralised.

Fourthly, the model also suggests that the SAU needs to have sufficient resources at its command so that it can respond efficiently and effectively to nullify any of the different types of errors the PSU can recognise.

One 'political' implication of the model is that the SAU will require the authority to terminate or suspend any of the activities that it controls without reference to other bodies. An unwanted error condition or disturbance could be of such a magnitude, when first perceived, that any additional time involved in seeking permission to apply corrective action could lead to a situation where the size of the hazard increases to such an extent that control is no longer possible.

1.5 Terminology

Yet another problem in risk management is that there are no general, publicly agreed definitions for much of the terminology used.[14,15] For example, the terms hazard and risk are frequently used as synonyms, yet this is inappropriate. Hazard refers to an unwanted phenomenon that can act upon the world and cause some form of harm if it should take place. Such an event could be a physical incident where the hazard eventuating might be something as simple as someone tripping over a rope and hurting their head in a fall. Or alternatively, the hazard could be a financial event which has an impact on an organisation, causing its share price to fall, as for example occurred when it became public knowledge that the National Westminster Bank had suffered 'massive trading losses which were possibly caused by a junior trader'.[16]

'Risk' is a concept often used to aid management decisions about hazards, and thus cannot act on the world itself. Technically, the term 'risk' is usually considered to consist of two components: first, a numerical probability (that is, a number between 1 and 0) that a particular hazard will eventuate; and secondly, a numerical estimate of the consequences that might result. When the two numbers are multiplied together the product is the numerical risk value associated with the specific hazard identified.

Thus, for example, if it was believed that a particular hazard had one chance in a million of occurring in a given period, and that, should the hazard be realised the financial consequences to the organisation or individual would be the loss of five million pounds sterling, then mathematically the risk would have a numerical value of five pounds sterling.

Therefore, it can be argued, that the term 'hazard' refers to the tangible whilst the term 'risk' refers to the abstract. Given that such confusion reigns it is perhaps not surprising that the management of risk does not receive the attention that it should in some quarters.

1.6 Benefits of risk management

The problems associated with risk management are not intractable and the benefits to be found in utilising risk management principles are significant. An illustration below in Figure 1.3, adapted from Allen,[17] attempts to represent the 'real world' difference between reactive and proactive risk management. In the diagram the number of unwanted incidents is plotted against time. The area under the curve represents the total loss to an organisation caused by the unwanted incidents. The top curve represents the situation with reactive feedback: that is, where the organisation waits until hazards are eventually reported and only then deals with them. The lower curve, on the other hand, depicts the reduction in losses that can be attained if organisations actively seek feedback on potential or actual hazards, allowing them to undertake remedial measures as soon as is practicable. The area between the two curves represents potentially preventable losses.

The way in which the benefits described in Figure 1.3 can be translated into financial performance is demonstrated in Figure 1.4. In this diagram an ideal typical benefit analysis is depicted. It is designed to demonstrate how the cost of controlled and uncontrolled risks varies over time. The top curve in the graph represents the predicted cost of risks with no control applied. For example, consider a warehouse that does not possess a water sprinkler system. If a fire should start it can be reasonably expected, given enough combustible material and oxygen, to burn until some action is taken to put it out. If no one perceives that there is a fire then it will continue to grow in size and will, given the right conditions, eventually burn the warehouse to the ground.

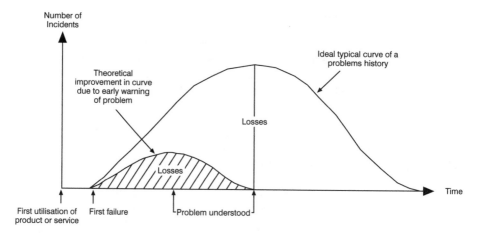

Figure 1.3 Development Period for a New Problem (adapted from Allen, D.E. (1992))

The bottom curve, however, shows that by investing in preventative control measures, the overall cost of risk is likely to be much smaller. If for instance, in the scenario described above, a water sprinkler system had been installed, when the fire started it would have been detected by the system and water applied to the affected area. Clearly if such action had taken place then the fire would have been put out much sooner and the damage, disruption and financial loss to the organisation would have been reduced.

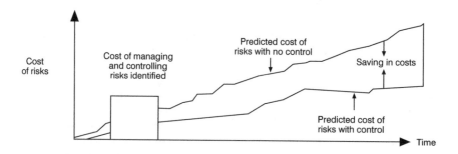

Figure 1.4 Ideal Typical Cost-Benefit Analysis

Comparing Figure 1.3 and Figure 1.4 in Figure 1.5 it can be observed that the cost of risk curve with no control in Figure 1.4 is comparable to the upper curve in Figure 1.3, and that the cost of risk with control in Figure 1.4 is analogous to the bottom curve in Figure 1.3. The area between the upper and lower curves in both diagrams represents the potential savings in costs to an organisation.

Figure 1.4 Ideal Typical Cost-Benefit Analysis

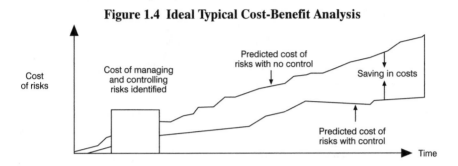

Figure 1.3 Development Period for a New Problem

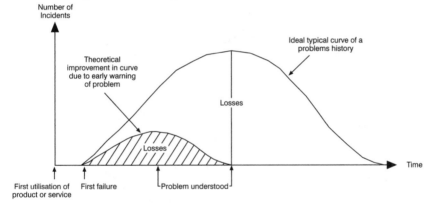

Figure 1.5 Comparison in Costs of Risks

1.7 Risk and quality management

Yet another way of understanding the utility of risk management is to consider the issue of quality management. The diagram in Figure 1.6 charts the way in which a number of quality defects can emanate from a product over time. The top curve represents the all too familiar reactive management approach to product quality problems. The bottom curve on the other hand illustrates the much smaller number of defects sustained when an organisation is actively engaged in proactive quality management. The difference in the areas covered by the two curves again indicates the potential savings in costs that can be made by preventing or reducing the number of defects.

We do not argue that quality management and risk management are the same: they cannot be because risk management also encompasses many other issues including those relating to safety. Although the two activities have similar aims, the prevention of unwanted incidents, the consequences for an organisation are likely to be significantly different where a customer is injured through a safety problem rather than inconvenienced through a quality problem. Clearly where an organisation causes injury it can lead to claims for compensation in the civil courts, prosecutions, fines or in extreme circumstances imprisonment. Typically, the worst outcome from a failure of product quality is a loss of custom.

However, the fact that risk management, cost-benefit analysis, and quality management can be modelled and related to each other in such similar ways supports the idea that the fundamental principles of risk management are, as argued earlier, inherent within all management systems. And consequently, the notion that any organisation that engages in proactive risk management can make significant improvements in both human and financial performance is a credible hypothesis.

The reason for this, in practical terms, is because if an organisation has in place a system which actively searches for, and analyses, data relating to risk management issues corporation wide, then any hazards should be identified on the positive slope of the lower curve in Figure 1.3.

Thus, once the hazard has been identified, investigation and remedial actions can be put in place much sooner than otherwise would have been the case.

Defect Prevention

Effect of Quality Planning

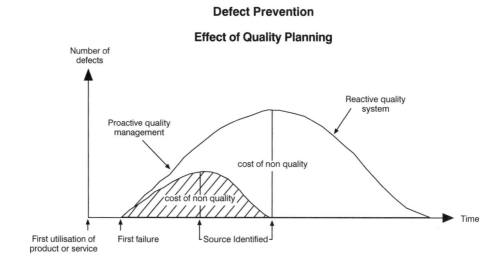

Figure 1.6 Quality Risk Management

1.8 Civil litigation

No win - no fee (contingency fee) services are now offered by lawyers in the UK and they look like they are here to stay. The face of civil litigation has changed for anyone who has, or believes they have, suffered a harm or an injustice by any organisation. Such a person can, subject to the strength of their case, have their day in court. Indeed, evidence of the popularity of such arrangements can be drawn from Terence Shaw who reports that:

> More than 10,000 accident victims who are ineligible for legal aid are
> believed to have made 'no win, no fee' agreements with their lawyers
> since the controversial scheme was introduced last summer.[18]

In 1996 there were a number of judgements made in the high court that will be of particular interest to those seeking compensation. The first is in relation to a medical condition known as Vibration White Finger (VWF). This particular industrial injury affects both the circulation and neurological function to the hand and is caused by prolonged exposure to vibration through the use of hand held tools. The vibration which the tools create can cause permanent damage and result in disability.

In a test case - *Armstrong and others v. British Coal Corporation* - Mr Justice Stephenson came to the conclusion that:

> British Coal ought to have known of the risks associated with tools used in the coal
> industry from at least 1 January, 1973, and that from January 1 1975, the company
> should have introduced a system of preventative steps including warnings and routine
> examinations of the miners.[19]

The second case concerns the families of children who had died of Creutzfeldt-Jakob Disease (CJD) following growth hormone treatment. High Court Judge Mr Justice Morland ruled that the Department of Health should have acted on warnings that some of the children might have contracted CJD as a result of the treatment and stopped using it after July 1977 when responsibility for its administration was transferred to it from the Medical Research Council.[20]

In both of these cases a high court judge has ruled that the organisation concerned was negligent because it failed to take appropriate action when evidence suggested that a medical problem existed. However, what is more important, as Fink points out in relation to VWF, is the:

> ...court's willingness to permit damages to accrue from a date by which a defendant is
> presumed to have had knowledge, despite the lack of direct evidence of such knowledge.
> In this way the 'Armstrong' decision has implications far beyond British Coal and VWF.[21]

Thus, these judgements should alert all organisations to the consequence of ignoring warning signs, for in each case, the problem simply got bigger and more expensive as more and more people were affected. It has been suggested that over 100,000 people may be able to win compensation in relation to VWF, while there are still some 200 actions pending with regard to CJD.

Given the decisions noted above, and the fact that the Law Society's Accident Line insurance scheme has reported that conditional fee arrangements are being signed at a rate of about 1,000 per month, it is clear that it is a dangerous strategy for any organisation to close its eyes and hope that a problem which has been identified will go away. The lessons to be drawn from the following chapters of this book reinforce this message but also offer organisations and individuals ways of minimising the incidence of such events.

Notes

1 Bannock, G. and Partners (1992) *Risk management, a boardroom issue for the 1990s*, Copies
   obtainable from Sedgwick Group Development Council, Sedgwick House, The Sedgwick Centre,
   London E1 8DX.

2 Otway, H. and Pahner, P. (1980) *Risk assessment in risk & chance; selected readings*, Dowie, J. and
   Lefrere, P. (eds.), Open University Press, p 157.

3 Douglas, M. and Wildavsky, A. (1982) *Risk and culture*, University of California Press.

4 Reid, S.G. (1992) *Acceptable risk in engineering safety*, Blockley, D. (ed.) McGraw-Hill, p 151.

5 Shrader-Frechette, K. (1991) Reductionist approaches to risk, in Mayo, D.G. and Hollander, R.D. (eds.), *Acceptable evidence: science and values in risk management*, Oxford University Press, p 220.

6 Slovic, P. (1992) Perception of risk: reflections on the psychometric paradigm, in Krimsky, S. and Golding, D. (eds.), *Social theories of risk*, Praeger, p 119.

7 Pidgeon, N., Hood, C., Jones, D., Turner, B.A and R. Gibson. (1992) *Risk perception, risk analysis, management and perceptions*, The Royal Society.

8 *Financial Times* (1995) 3 March.

9 Symes, A. (1993) Handling risk perception, *Occupational Health Review*, September/October, p 31.

10 Seccombe, I. and Ball, J. (1992) *Back injured nurses*, Institute of Manpower Studies, November.

11 Toft, B. (1996) Limits to the mathematical modelling of disasters, in Hood, C. and Jones, D.K.C. (eds.) *Accident and design: contemporary debates in risk management*, UCL Press.

12 Toft, B. and Reynolds, S. (1994) *Learning from disasters: a management approach*, 1st edition, Butterworth-Heinemenn.

13 Hidden, A. (1989) *Investigation into the Clapham Junction railway accident*, Cm 820, HMSO.

14 Health and Safety Executive (1995) *Generic terms and concepts in the assessment and regulation of industrial risk,* Health and safety executive, DDE2,1.

15 Boyle, T. (1997) A fresh look at risk assessment, *The Health and Safety Practitioner*, February.

16 Ladbury, A. (1997) *Risk management failings highlighted at Natwest*, Business Risk, Association of Insurance and Risk Managers in Industry and Commerce, Spring.

17 Allen, D.E. (1992) *The Role of regulation and codes, in engineering safety*, D. Blockley (ed.), McGraw-Hill.

18 Shaw, T. (1997) *Daily Telegraph*, 5 June.

19 Taylor, R. (1996) *Financial Times*, 16 January.

20 *Financial Times*, (1996) 20/21 July.

21 Fink, S. (1996) The presumption of knowledge, *The Health and Safety Practitioner*, July.

Chapter 2

Disasters as systems failures

Introduction

In this chapter it is argued that many of the popular ideas regarding the underlying causes of technological disasters are myths. Examples include the views that such events are the product of divine wrath, or are solely technical in nature. The former suggests we cannot learn from these events since divine intervention is inexplicable, while the latter suggests that an engineering solution will of itself be sufficient to prevent a recurrence of the incident. However, much research suggests that the underlying causes of catastrophes are far more complex than the simple explanations generated by such beliefs. Subsequent analysis of these events reveals that their underlying mechanisms invariably have organisational and social dimensions, while technological factors are sometimes, but not always, present. Utilising the theoretical framework of systems theory, and the concept of organisations as socio-technical systems, analysis allows technological disasters to be more appropriately understood as the result of human rather than divine actions. Similarly, this mode of analysis flags up the more complex socio-technical nature of these events as opposed to the exclusively technical.

2.1 From act of God to technical failure?

Seldom does a month go by without a major catastrophe occurring somewhere in the world with a consequent tragic toll of death, injury, and property damage. These disasters occur in people's homes, transport systems, communities, and especially their workplaces. Unfortunately, in popular descriptions of a disaster words like freak, unforeseen, technical failure, act of God ... are often used to explain the reasons for the incident. Such language fosters the beliefs that disasters can indeed be explained as solely the result of some technical defect, or as an act of divine caprice.

An early example of the latter explanation occurred immediately after the collapse of the railway bridge over the River Tay in Scotland during a storm on the evening of 28 December 1879. Seventy-five people lost their lives in the disaster. The following Sunday Dr Begg announced from his pulpit:

> If there is one voice louder than others in this terrible event it is that of God! Determined to guard his Sabbath with jealous care, God does not afflict except with good cause. The Sabbath of God has been dreadfully profaned by our great public companies. These wicked people are actually going to have the audacity to rebuild this bridge. Is it not awful to think that they [the passengers] must have been carried away when they were transgressing the law of God.

A more recent example highlighting the propensity to present a solely technical explanation following a major disaster occurred immediately after the *Challenger* Space Shuttle accident in January 1986. An 'O' ring, intended to prevent hot gases escaping from a booster rocket, was identified as having failed and almost exclusively blamed for creating the disaster.[1] This apparent tendency to look for simple causal solutions when disaster strikes may be one of the chief reasons why issues of safety, particularly in relation to technological issues, often appear to be almost solely shaped by the technical concerns of the engineering community.

While the 'O' ring failure in the *Challenger* accident was correctly identified as being the physical reason for the catastrophe occurring on that day, there were a number of other equally important although less tangible factors involving policy, procedures and practice. These were subsequently brought to light in the 1986 report of the US House of Representatives' investigation. Similarly, in the wake of other recent large-scale accidents – particularly those at Three Mile Island, Chernobyl, King's Cross, Piper Alpha, and the Clapham Junction rail collision – engineers or technical experts were put in the spotlight by the media. Consequently the press initially speculated about the possible technical or engineering malfunctions that might have been responsible for the incident. In each case, however, investigations revealed that there were other equally important organisational and social factors at work which had played crucial roles in the creation of these events.

The inherent power of these factors to create an environment in which a catastrophe can readily occur can be clearly observed in the capsizing of the Zeebrugge cross-channel ferry; the sinking of the *Marchioness* pleasure boat on the river Thames; the loss of life in the crush of spectators in the Hillsborough football ground, and in the fire in the football stand at Bradford. In each of these disasters no physical malfunction of any equipment, plant or premises was found to have taken place. On the contrary, these were occasions where a mixture of human, organisational and social pathologies were found to be entirely responsible for the incidents occurring.[2]

2.2 Socio-technical systems

Since Dr Begg's emotional sermon the study of the disaster causation has advanced a long way. It is clear that it is no longer appropriate to consider disasters in fatalistic or solely technical terms, but that a much wider set of circumstances must be addressed. Turner rightly argues that in the search for some general principles to aid our understanding of disasters:

> ...it is better to think of the problem of understanding disasters as a 'socio-technical' problem with social organisation and technical processes interacting to produce the phenomena to be studied.[3]

Indeed the reports of the public inquiries into the accidents noted above support Turner's call for a change in the way disasters are perceived. Similar conclusions are drawn by many other researchers in the area, all of whom argue that the aetiology of disasters are typically comprised of a complex combination of technical, individual, group, organisational and social factors. Disasters may be described as resulting from the breakdown of what, in the aftermath of Chernobyl, was termed by one British government minister, an organisation's 'safety culture.'[4]

2.3 Safety culture

The idea of safety culture referred to above is just one facet of the concept of culture, widely used in social science. A multiplicity of definitions of the term culture are available, but here it is regarded as the collection of beliefs, norms, attitudes, roles and practices of a given group, organisation, institution and society. In its most general sense culture refers to the array of systems of meaning through which a given people understand the world. Such a system specifies what is important to them, and explains their relationship to matters of life and death, family and friends, work and danger. It is possible to think of a hierarchy of cultures, with the cultures of small groups of workers, of departments, of divisions and organisations being nested successively within one another. The organisational culture can then be considered as being located within a national or international framework.

A culture is also a set of assumptions and practices which permit beliefs about topics such as danger and safety to be constructed. A culture is created and recreated as members of it repeatedly behave in ways which seem to them to be the natural, obvious and unquestionable way of acting. Safety culture can be defined as those sets of norms, roles, beliefs, attitudes and social and technical practices within an organisation which are concerned with minimising the exposure of individuals to conditions considered to be dangerous. (This discussion of safety culture draws heavily on Pidgeon et al[5].) Waring draws the conclusion from his research that:

culture is not a 'thing' but a complex and dynamic property of human activity systems.[6]

Similar conclusions are drawn by Turner et al.[7]

The following examples highlight the role which safety culture has played in recent disasters. Lord Cullen noted in his report on the Piper Alpha disaster that:

It is essential to create a corporate atmosphere or culture in which safety is understood to be, and is accepted as, the number one priority.[8]

Similarly, Desmond Fennell QC, in his report on the King's Cross inquiry, when discussing the organisational aspects of the disaster observed that:

London Underground has accepted that a cultural change is required throughout the organisation.[9]

Again, John Hayes in his report on river safety concluded that:

Our Enquiry has shown that a safety culture does not come naturally to those who work the river for reward. They need to be led into this culture by their owners and by the regulators.[10]

Clearly an organisation's safety culture does not spring into existence overnight as a mature phenomenon. It takes time for the complex sets of individual and collective perceptions to develop and coalesce into a system of commonly shared values. Therefore it is perhaps hardly surprising that Johnson argues that the evidence suggests that an organisation's culture:

...is highly resistant to change; leaders who try to change it do not realise how the manner in which they pursue change confirms the existing culture.[11]

Turner et al. similarly note that:

...while an appropriate corporate culture is crucial to good performance in most industrial situations, such cultures are both more subtle and more resistant to manipulation than many had at first assumed. This is not to say, as did the OECD Nuclear Energy Agency, that 'attitudes cannot be regulated', but rather to indicate the difficulty of achieving such regulation by imposition from the upper levels of an organisation.[12]

2.4 Systems concepts

The idea that many systems have underlying similarities and possess common properties was developed by the biologist von Bertalanffy in the 1920s and 1930s:

> It did not matter whether a particular system was biological, sociological or mechanical in origin, it could display the same (or essentially similar) properties, if it was in fact the same basic kind of system.[13]

From this concept the idea of organisational isomorphism was developed, discussed further in Chapter 5. Two corollaries can be inferred from the above hypothesis. First, any failure that occurs in one system will have a propensity to recur in another 'like' system for similar reasons. Second, although two particular systems may appear to be completely different, if they possess the same or similar underlying component parts or procedures then they will both be open to common modes of failure.

Support for these arguments can be found in the work of many researchers.[14-20] A recent comprehensive review of the literature on man-made disasters was carried out by Horlick-Jones who suggests that work in the area 'increasingly reflects the realisation that disasters are systems failure'.[21]

Two examples illustrate the organisational isomorphism concept. First, the similarity between two fires: one at the Iroquois Theatre, Chicago in December 1903, the other at the Coconut Grove Nightclub, Boston in November 1942. In both cases the decorative fabrics of the interiors were highly inflammable, and exits were either locked or had not been provided. Both venues were overcrowded and in neither establishment had staff been trained to deal with emergencies such as fire. This pattern is not unfamiliar in places of entertainment and the list of cases could be extended without difficulty.

A second example shows an underlying similarity in the failure of systems of two apparently disparate types – a bridge and nuclear power plant. In October 1970 the West Gate Bridge near Melbourne, Australia collapsed after a structural failure. It was later demonstrated that this was in part caused by the unsatisfactory nature of a computer program used to calculate the size and disposition of steel required for the bridge's construction. In March 1979 it was reported that five nuclear power plants in the USA had been closed down due to an apparently 'simple mathematical error' discovered in a computer program used to design the reactor cooling system.[22] In both these quite different systems a broadly similar problem – defective design software – led to undesirable results.

It is argued here that disasters are rarely the result of technical factors alone; rather they are failures of complex systems. The underlying philosophy of this book is that socio-technical failures are not the result of divine caprice, nor of a set of random chance events which are not likely to recur, nor simple technical failures. Rather, disasters are incidents created by people operating within complex systems. These incidents can be analysed, and the lessons learned applied to prevent or reduce the chance of similar events recurring. The evidence suggests that where lessons are not learned similar accidents can and do recur.

2.5 A failure of hindsight

Several tragic examples highlight this failure to learn. The fire in a stand at Bradford City Football Club on 11 May 1985 cost the lives of more than 56 people among the crowd that had turned up to watch a football match. Many years earlier, in August 1969, the Fire

Protection Association had published an article in its journal detailing several fires in football stands like the one at Bradford and warning of fire risks associated with such stands.[23] If this information had been brought to the attention of the Bradford Football Club staff, and acted upon, then the tragedy there would never have occurred.

Another example is that of the loss of coolant water at the Crystal River nuclear plant in Florida USA on 26 February 1980. An identical valve to that which malfunctioned at Three Mile Island in March 1979 stuck open and flooded the reactor basement with 190,000 litres of highly radioactive water. In this case the Nuclear Regulatory Commission, which had conducted the inquiry into the Three Mile Island incident, had permitted the plant operators two exemptions from the recommendations which they themselves had suggested. According to some nuclear scientists those two exemptions allowed the Crystal River incident to develop into a major emergency.[24]

An identical valve to that which stuck open at both Three Mile Island and Crystal River malfunctioned again in precisely the same way at Davis-Besse in Ohio on 9 June 1985. Fortunately on this occasion an operator noticed the drop in pressure and prevented the reactor from entering a dangerous condition.[25] In the period between the two incidents, a report was published in Britain on an incident at the Heysham nuclear power station, Lancashire, which discussed mistakes that mirrored Three Mile Island.[26]

A further example of the failure of some organisations to learn from hindsight is clearly illustrated by the circumstances within British Rail prior to the Clapham Junction railway disaster in December 1988 in which 39 people died, and 500 were injured, 69 of them seriously. The Clapham Junction Public Inquiry concluded that:

> The accident need not have happened if previous WSFs (Wrong-sided Signal Failures, that is a signal failure that leaves the system in a dangerous condition) had been investigated in order to establish the lessons to be learned and if those lessons had been effectively taught to relevant staff. The lessons from the Oxted incidents in November 1985 and the Queenstown Road incident in June 1988, should have prevented the tragic combination of circumstances which led to the Clapham Junction accident. BR must therefore take every possible lesson from this accident and ensure that appropriate action is taken and regularly monitored and audited.[27]

These cases relate to the failure of organisations to absorb information from earlier disasters, but other examples show cases of organisations failing to respond appropriately to the risk of a major accident even after being forewarned.

At Aberfan in South Wales 144 people died in October 1966 in a slippage of a waste tip at a mine pithead. There had been many complaints from the local population about the dangers that the tip posed, especially after a number of previous slips on nearby tips.

A British Rail sleeping car burned at Taunton in 1978 after bed linen stored in a corridor caught fire and set the rest of the carriage ablaze. Twelve lives were lost. It was reported at the inquest after the tragedy that:

> British Rail had a warning five years ago that bed linen left on the sleeping car heater was a source of danger... There was an inquiry then after linen caught fire on a Glagow-Euston train, but British Rail issued no general warning.[28]

Another example is that of the King's Cross underground station fire in London in November 1987 in which 31 people lost their lives. Prior to this disaster there had been fire incidents at other London underground railway stations which had been the subject of internal inquiries.[29,30]

2.6 Organisational learning – helping change safety culture

Why does this failure of hindsight occur? Why is the corporate safety culture not always the number one priority? How can that culture be changed? Unfortunately there is a tendency for some managers to assume that culture change can be achieved quickly by management decree or legislative prescription, as if changing a culture is like changing a formal management procedure. Policy and framework are necessary, but since culture is both a product and a moulder of people the emphasis should be on 'organisational learning' which recognises a long-term process. Organisational learning may be defined as a cumulative, reflective and saturating process through which all personnel within organisations learn to understand and continually reinterpret the world in which they work by means of the organisational experiences to which they are exposed.

Attempts to change the safety culture by higher management decree have no guarantee of a successful outcome. Top management may be dedicated to the task and commit resources, but research has shown that this is not always enough. Staff changes, suspicion from lower levels within the organisation, efforts tailing off after initial enthusiasm, are some of the factors that may thwart success.

Legislative prescription carries problems as well. Simply passing safety legislation does not automatically mean that measures are effectively implemented in practice. Lord Cullen wrote in the Piper Alpha report:

> No amount of detailed regulation for safety improvements could make up for the deficiencies in the way that safety is managed by operators.[31]

Prescription might even have the opposite effect to that intended and increase risk rather than lower it. The report into the 1975 Fairfield Home fire stated:

> ...some architects tend to rely on regulations and available guides, rather than on their own understanding of the principles, to design against fire risk.[32]

The Piper Alpha inquiry report recommended extension of the concept of the 'safety case' to the UK's North Sea offshore sector, highlighting the legislative trend away from detailed prescription. Safety cases, a development of the EC's Seveso directive, require organisations working in hazardous industries to formally assess their risks and devise methods of lowering those risks. This contrasts with having safety measures fixed in stone by prescriptive legislation. The provisions of the Seveso directive came out of acknowledgement of the problems of prescription. At a more detailed operating level, organisational personnel often find that prescriptive safety measures and procedures can become routine and boring after a while. As a consequence habits are formed which break or circumvent the rules of the organisation.

Neither decree, nor prescription, nor technical approaches on their own are sufficient to effect permanent change in the safety culture of organisations. Given the difficulty of changing the culture quickly a sustained commitment must be made to change. Organisations are dynamic, complex human activity systems. Both they and the environment in which they operate are continually evolving. Therefore, if organisations are to be as safe as reasonably practicable they need to learn from their own experiences and, where appropriate, the experiences of others. As well as decree and prescription, positive changes to an organisation's safety culture require 'organisational learning' and sustained commitment to safety issues.

2.7 Systems failure models

In order to more fully understand some of the general organisational features and processes found before, during and after a catastrophic incident a model of system failure and cultural readjustment has been developed. The model, which provides a context for the present work, is based upon Turner's six-stage Disaster Sequence Model[33] and an analysis of case studies carried out for the research on which this book is based.

2.7.1 Systems failure and cultural readjustment model (SFCRM, Figure 2.1)

At the most general level the model consists of three separate but interrelated parts. The first part includes the incubation period of actions and events prior to a disastrous system failure. Turner developed the notion of incubation, where latent defects build up within the system. The second part includes the event triggering the disaster, the disaster itself, and the immediate aftermath of rescue and salvage. The third part includes the learning process – investigation and inquiry and the production of reports and recommendations. Of fundamental importance is a feedback channel from this third, learning part back to the first, incubation part. This negative feedback loop, external to any organisation involved, carries the lessons of the disaster back to the operational socio-technical environment, allows organisational learning to take place, and so helps prevent similar disasters. The model attempts to capture graphically some of the fundamental system elements that appear to be present in the generation of a 'disaster cycle'.

2.7.2 The SFCRM model in detail

At a more detailed level the SFCRM model may be described in seven stages. Before discussing these it should be noted that the term 'environment' in the context of this model should be interpreted in a catholic manner. It refers to the external conditions in which people live and work. For example, an airline pilot's working environment would be the aircraft in which he or she flies, and any of the areas of the airport or airline offices used in connection with that employment. The pilot's living environment refers to hotels, countries, places visited as well as his or her permanent residence.

Part 1

Stage 1: The incubation period
The cycle is assumed to begin with someone perceiving a need for a change in the environment. The perceived need might be a new bridge or power station, or a change in a chemical process plant, or the building of a new factory. It might be something as simple as the refurbishment of an old hospital ward.

Once the need has been recognised and sufficient financial backing made available the project will be passed to some form of 'design system'. The designer translates the required environmental change into working plans. From the designer the working plans are given to a 'building and construction system'. The 'builder' carries out the work necessary for the desired change in the environment.

This change may create a completely new environment – say, a new chemical plant on a new site – or a modification of the existing environment – say, the addition of a new wing to an existing hospital. Whichever of these, the result is a new physical and organisational entity.

Stage 2: The operational socio-technical system
When the changes to the environment are completed personnel begin to work within its confines and a new socio-technical system can be said to be operational.

Part 2

Stage 3: The precipitating event
The socio-technical system operates, possibly sufferingfrom inherent latent system failures which build up until some event occurs that precipitates or triggers an incident, accident or disaster.

Stage 4: The disaster itself

Stage 5: Rescue and salvage
Occurs immediately after a disaster. This stage involves initial attempts to mitigate and recover from the effects of the disaster.

Part 3

Stage 6: Inquiry and report
Following a disaster there is usually an internal investigation by the directly affected organisations to determine causes. Typically a disaster raises public, media and official concern. Governments usually require a Public Inquiry to allay public fears. Forensic evidence, eye witness reports, expert witness statements are collected and all the evidence is subjected to public scrutiny during the course of the inquiry. At the end of the inquiry a report on the investigation is published and submitted, in Britain, to the Secretary of State of relevant government departments, and to the organisations involved. The report contains recommendations that the inquiry team considers relevant to prevent a repetition of the incident.

Stage 7: Feedback
The receipt and implementation of the recommendations from the inquiry by the organisation concerned and by organisations exposed to the same or similar risks. Frequently interim recommendations will have already been made after initial investigation and will have immediately been fed back to the organisation concerned before the main inquiry has finished its deliberations.

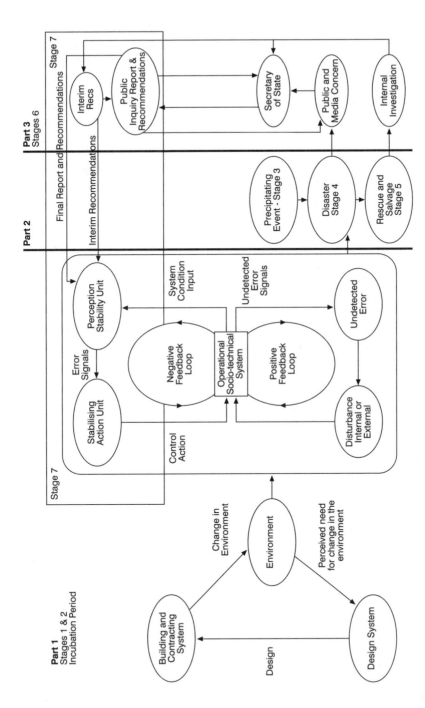

Figure 2.1 Systems failure and cultural readjustment model

2.7.3 Turner's model

There is a strong correspondence between the stages of the SFCRM model described above and the Disaster Sequence Model developed by Turner. The seventh stage – feedback and cultural readjustment – is implicit rather than explicit in Turner's model, which he describes as follows:

Stage I – notionally normal starting points:
(a) initially culturally accepted beliefs about the world and its hazards;
(b) associated precautionary norms set out in laws, codes of practice, mores and folkways.

Stage II – the incubation period: the accumulation of an unnoticed set of events which are at odds with the accepted beliefs about hazards and the norms for their avoidance.

Stage III – precipitating event: forces itself to the attention and transforms the general perceptions of Stage II.

Stage IV – onset: the immediate consequence of the collapse of cultural precautions becomes apparent.

Stage V – rescue and salvage: first stage adjustment – the immediate post-collapse situation is recognised in ad hoc adjustments which permit the work of rescue and salvage to be started.

Stage VI – full cultural readjustment: an inquiry or assessment is carried out and beliefs and precautionary norms are adjusted to fit the newly gained understanding of the world.'[34]

Stage VI of Turner's model comprises both the 'inquiry' and change in 'beliefs and precautionary norms'. It suggests that knowledge gained from inquiries is absorbed into the culture of organisations. Implicit within the model is the implementation of inquiry recommendations by organisations concerned and organisations exposed to the same or similar hazards. As will be argued later this is not always the case.

2.8 Feedback within the socio-technical system – positive and negative

The second stage of part 1 of the SFCRM model shows two feedback loops around the operational socio-technical system. The upper loop is a negative feedback loop and is regarded as a 'normal' control action which attempts to keep the organisation's operational parameters within an envelope defined for the system. If because of internal or external disturbance, any of the system's parameters goes out of bounds of this envelope this 'instability' should be perceived, a negative error signal generated (that is an error signal opposing the change in parameter), and then stabilising action should take place to bring the system back within bounds.

The lower loop is a positive feedback of a 'dysfunctional' action in which an internal or external disturbance is undetected and therefore uncorrected. This disturbance may give rise to dormant failures and errors, which build up within the socio-technical system until a precipitating event leads to disaster. This behaviour is typical of what takes place during the incubation period in part 1 of the model.

During normal operations the organisation will be endeavouring to produce some form of stable output, for instance the construction of a bridge or the production of chemicals. However, if there is a disturbance or perturbation from normal operation of an organisation's processes this should be identified by a 'PSU.' This could be anything from an operator in a plant noticing that an instrument reading is outside the allowed tolerance, to a ground proximity warning signal in an aircraft cockpit, to a computer controlling a chemical process plant sounding a warning. Upon identifying the abnormal operation the perception unit will pass control to a 'SAU' which acts to reduce the effect of the disturbance and bring the systems parameters back within bounds. The term negative feedback is used since the control action is always in opposition to the perceived disturbance. This process is depicted in the upper loop of figure 2.1.

If during normal operations a dysfunctional disturbance is not perceived the effects of the disturbance may remain dormant within the system, or worse cause some positive feedback. In this latter case the disturbance may put the socio-technical system into such a state that an even larger undetected disturbance is generated. Here the term positive feedback is used because the system action is in the same direction as the original disturbance and adds to its effects.

An example of the damaging effect of positive feedback can be seen in the spread of rule breaking within an organisation. If people begin to break rules and this 'disturbance' to the socio-technical system is perceived then control action (some form of sanction) can be applied and the rule breaking diminished. If the disturbance is not perceived, or if it is perceived and no control action takes place, then more and more rules may be broken by more and more people until in an unforgiving environment – say, a coal mine or oil platform – it is quite likely that at some point a catastrophic systems failure would occur. Positive feedback loops should always be treated with extreme caution, even when they are required as is the case in certain industrial situations.

It should also be noted that a perturbation could be so large that even though perceived the amount of negative control action at the organisation's disposal is less than required to restore stability. If this is the case, then again some form of organisational system failure would be sustained.

2.9 Negative feedback

From the above can be seen the importance of the negative feedback process – disturbance detection, stability perception and stabilising action – to the continued normal operation of an organisation. It can be seen that stage 7 of the SFCRM model – the public inquiry and report process – endeavours to provide a negative feedback loop external to the organisation. This external loop feeds its information into the organisation's own internal negative feedback loop so as to attempt to prevent recurrence of similar incidents.

At the most detailed level of analysis it is obvious that no specific accident occurs twice. Each individual disaster is unique – for instance the fire at the Summerland Leisure Centre on the Isle of Man cannot occur again in its original specific detail. Most of the building was destroyed in the fire and the people involved will never be in the same place at the same time again. However, at another level of analysis it is possible for a Summerland type of fire to occur. For example, as we have already seen, if any crowded place of public entertainment catches fire and attempts to control the fire are unsuccessful, problems of escape are posed. If staff have not been trained in emergency procedures, if the evacuation of patrons is delayed, if exit doors are locked to prevent illegal entry, and if the contents of the building are highly flammable, then the likelihood of a similar disaster occurring is extremely high.

As noted earlier, in Turner's model it is assumed that once the causes of disaster have been determined by public inquiry, and recommendations made to prevent a recurrence of the event, a change in the safety culture of the organisation concerned and the industry will take place. That is, the information gained as a result of a disastrous failure will be absorbed into the prevailing beliefs and practices within organisations to avoid future similar incidents. However, while it can be argued that cultural change does take place in some organisations that have suffered a disaster, there are plenty of examples where lessons learned by one organisation are not implemented by others in the same or related fields. These organisations are therefore at risk of the same kind of failure for the same or similar reasons.

Models such as Turner's tacitly assume that the organisations involved will implement all of an inquiry's recommendations, both interim and final. There are, however, two problems associated with this assumption. First, while the inquiry may have resulted in increased awareness of the hazards and made recommendations to deal with them, if the organisation involved does not implement any control action then the situation that existed before the disaster will continue. Second, if only some of the recommendations are implemented an organisation may only be marginally safer than it was before the disaster occurred, and therefore may still not be properly protected against a recurrence of the event.

Learning the lessons of disasters is a two fold process: perception and implementation. It is not sufficient just to perceive the events which create disasters. Effective learning requires immediate implementation of those lessons by using control action within organisations. However, sometimes it is not possible for organisations to act in such an optimal way. This is discussed further in Chapter 6.

It also appears to be tacitly assumed by both the public and those responsible for investigating a disaster that other organisations which share similar hazards will also pick up and implement the recommendations made by the inquiry. However, while it would appear that organisations other than those involved in a particular incident do register that a disaster has taken place in their industry, sadly, they often do not appear to incorporate the findings of the inquiry into their organisation unless legislation is passed to that effect.

The notion that the recommendations that emerge after public inquiries always lead to 'full cultural readjustment' within organisations is at odds with at least some of the empirical evidence. In the cases noted above there was a failure to use the hindsight gained from previous accidents and incidents to prevent further disasters occurring.

Hindsight is without doubt one of our most important and most costly information sources, both in terms of lives and capital expenditure lost. We should attempt to gain as great an understanding as possible from such information when it is presented to us. The information should be used as effectively as possible so that the benefits gained are maximised and that any further unnecessary 'costs' in the form of future disasters are kept to a minimum.

An extensive review of the literature on organisational learning suggests that a central concern of such research is the identification of conscious and explicit attempts by organisations to learn from positive and successful events.[35] This may involve internal experimentation, imitating strategies developed by other organisations, or utilising third-party consultants.

The research on which this book is based tried to identify positive behaviour that has evolved in organisations endeavouring to explicitly learn from negative events, that is disasters. This is where, in mainstream research into organisational learning strategies, the analysis appears to stop.

The research looked at the learning processes of organisations which become involved in public inquiries through their being associated in some way with a socio-technical failure. The rationale behind this line of research activity was that public inquiries do present unique opportunities for organisations to learn from their mistakes and change their safety culture. Indeed it is one of the stated functions of public inquiries to try to prevent similar events recurring. Therefore the research was designed to assist in our understanding of this reporting and learning process which takes place after such events, in the hope that it may eventually be possible to design organisational structures which will help reduce accidents.

Notes

1 For example see *The Guardian* (1986) 1 February.

2 Toft, B. (1992) Changing a safety culture: decree, prescription or learning? Paper presented at a IRS training conference on risk management and safety culture, London Business School, London, April.

3 Turner, B.A. (1978) *Man-Made Disasters*, Wykeham, London.

4 Cited in ATOM (1987).

5 Pidgeon, N., Turner, B.A., Toft, B. and Blockley, D. (1992) Hazard management and safety culture. *Hazard management and emergency planning: perspectives on Britain* Parker, D. and Handmer, J. (ed.), James X James, London.

6 Waring, A. (1992) Developing a safety culture, *The Safety & Health Practitioner*, April.

7 Turner, B.A., Pidgeon, N. Blockley, D. and Toft, B. (1989) Safety culture: its importance in future risk management, Position Paper for The Second World Bank Workshop Safety Control and Risk Management, Karlstad, Sweden, 6-9 November.

8 Cullen, The Hon Lord. (1990) *The Public Inquiry into the Piper Alpha Disaster*, HMSO, London.

9 Fennel, D. (1988) *Investigation into the King's Cross underground fire*, HMSO, London.

10 Hayes, J. (1992) *Report of the enquiry into river safety*, Cm 1991, HMSO, London.

11 Johnson, B.B. (1992) Risk culture research: some cautions. *Journal of Cross-Cultural Psychology*, 22, March.

12 As note 7.

13 As cited in Beishon, J. (1980) Introduction to systems thinking and organisation. *Unit 1/2 Systems Organisation: The Management of Complexity*, Course T243, Block 1, Open University Press, Milton Keynes.

14 Ackoff, R.L. (1980) The systems revolution. *Organisations as systems* Lockett, M. and Spear, R. (eds.), Open University Press, Milton Keynes.

15 Buckley, W. (1980) Sociology and modern systems theory. *In Organisations as Systems*, op. cit.

16 Checkland, P.B. (1981) *Systems thinking, systems practice*, Wiley, Chichester.

17 Bignell, V. and Fortune, J. (1984) *Understanding systems failures*, Manchester University Press in association with the Open University.

18 Toft, B. (1984) Dissertation for the awarding of a BA(Hons) degree in independent studies, University of Lancaster.

19 Beer, S. (1985) *Diagnosing the systems for organisations*, Wiley, Chichester.

20 Kletz, T.A. (1988) *Learning from accidents in industry*, Butterworths.

21 Horlick-Jones, T. (1990) *Acts of God? An investigation into disasters*, Association of London Authorities.

22 As reported in *New Scientist* (1979) 22 March.

23 *FPA Journal* (1969) No. 83, August.

24 As reported in *New Scientist* (1980) 6 March.

25 As reported in *New Scientist* (1985) 11 July.

26 As reported in *New Scientist* (1984) 23 November.

27 Hidden, A. (1989) *Investigation into Clapham Junction railway accident*, Cm 820, HMSO, London.

28 As reported in *The Times* (1978) 18 August.

29 As reported in *New Scientist* (1985) 21 February.

30 As reported in *New Scientist* (1988) 25 February.

31 Cullen, The Lord, op. cit.

32 *Report of the Committee of Inquiry into the fire at Fairfield Home, Edwalton, Nottinghamshire on 15th December 1974* (1975) HMSO, London.

33 Turner, op. cit.

34 ibid.

35 Levitt, B. and March, G. (1988) Organisational learning. *Annual reviews in sociology*, 14: 319-340.

Chapter 3

Methodology

Introduction

This chapter describes the underlying methodological assumptions, and the analytical techniques adopted in this research. The research underlying this book is based on the collection and analysis of data relating to how organisations attempt to learn from socio-technical disasters. Data were collected from organisations represented at nineteen public inquiries into disasters that occurred in the UK between 1965 and 1978. Information was gathered from face-to-face interviews, letter correspondence and documentation, both published and unpublished. Multiple data sources were utilised where possible. The collected data were then analysed using 'grounded theory'.

3.1 Methodological perspective

While there are many unresolved arguments within the philosophy of the social sciences, including the 'mind and body problem' and 'nature versus nurture', for the purposes of this research these theoretical difficulties are set aside. The methodology employed in this research is based on two underlying assumptions.

a) There is a 'real world' in which people have experiences, and these experiences can be shared with others through some form of communication. This view is supported by Brown and Sime who suggest:

> ...that people can and do comment on their experiences, and that these commentaries are acceptable as scientific data.[1]

b) People typically tell the truth more often than not.

This view of the communication of information between interviewer and respondent is based on Davidson's 'Principle of Charity'.[2] This does not imply that what respondents say should be cast in stone and accepted for all time as the truth. Rather it suggests that, in the absence of any evidence to contradict what respondents profess to be the case, it should be assumed that they are not deliberately lying or distorting the 'facts' as these appear to themselves.

3.2 Choice of methodology

The methodology chosen for this research comprised three elements – case studies of selected disasters, the collection of data from multiple sources utilising the notion of triangulation, and data analysis using grounded theory.

The validity of the case study approach is supported by Goode and Hatt, who argue that case studies are:

> A way of organising social data so as to preserve the unitary character of the social object being studied...an approach which views any social unit as a whole.[3]

3.2.1 Data collection

Data collection methods included: semi-structured interviews with crucially placed organisational personnel; use of published and unpublished documents; and personal correspondence with organisations. Collection of data from multiple sources allowed use of the methodology termed 'triangulation'. The central idea here is that, according to Webb:

> Once a proposition has been confirmed by two or more independent measurement processes, the uncertainty of its interpretation is greatly reduced.'[4]

Arber points out the benefit of multiple methods in gathering data to develop and test a theory:

> No single method permits a researcher to develop causal propositions free of plausible rival interpretations. All methods have weaknesses but there can be 'strength in converging weakness' by confronting theories with a series of complementary methods of testing, so increasing the validity of interpretations.[5]

The methodology adopted generates evidence from several different perspectives. This is particularly useful when dealing with events which occurred a considerable time ago, since respondents do forget events and they reconstruct their interpretations of what occurred.

Semi-structured interviews, where used, ensured that the same questions were asked of each respondent, but had enough flexibility to allow different responses to those questions.

3.2.2 Analysis

It was decided that the most appropriate method of analysis for the collected quantitative data was 'grounded theory'. That is, the research produces results in which the theory is grounded in the data. Support for this position comes from Martin and Turner who note:

> Grounded theory is particularly well suited to dealing with qualitative data of the kind gathered from semi-structured or unstructured interviews, from case study material or certain kinds of documentary sources... The grounded theory approach offers the researcher a strategy for sifting and analysing material of this kind.[6]

A further consideration behind the choice of this particular analytical technique is the authors' firm belief that theory should be formulated out of practice and that the two should be inextricably linked together. As Turner notes:

> A thesis is tied together by an argument. This argument should run through the thesis like a rope. It should be strong and continuous... When any piece of the thesis is seized, it should pull upon the rope of the argument.[7]

Arguments in favour of quantitative methods were not forgotten, but it was decided that quantitative techniques were unsuitable for the interviewer-respondent question and answer method utilised in this research. Little of the relevant response material was in quantifiable form, and as Marshall notes:

> researchers must match research methods with research questions.[8]

3.3 Operational methodology

Four broad criteria were used to select disasters:

a) The disaster should have occurred sufficiently long ago so that the sensitivities of the respondents were now low enough to enable them to talk freely without getting too emotionally upset.

b) The disaster should have occurred recently enough so that:
    i) the people involved would still remember the events with some clarity;
    ii) documentation of what occurred might still be available;
    iii) it was likely that the people involved would still be with the organisation, or traceable if they had changed employer or retired.

c) There should be no litigation pending. This is because if litigation is involved few people would be willing to talk about the event in case they revealed something which may be construed as admitting fault or liability.

d) The disasters should have occurred in the United Kingdom to organisations under British management. The lessons learned, and the methods and processes used to implement them, would then reflect one single culture. This strategy was adopted because it was thought that different nationalities might well learn and implement their lessons in completely different ways and as a result might prevent the drawing of common patterns from the complete data set.

Nineteen disasters, fulfilling all the criteria mentioned above, were selected. Of these nineteen, five disasters were selected to be the focus of the interview programme.

Organisations represented at the public inquiry into each of the nineteen disasters were identified. Only organisations represented by counsel at inquiries were sought, for two reasons:

a) as they were represented they were unlikely to have been at the inquiry solely as expert witnesses or in an advisory capacity, but were there because of direct involvement in the disaster;

b) evidence given by represented organisations is more likely to give an organisation-wide perspective on the disaster rather than an individual perspective.

Seventy-nine organisations were involved in the public inquiries into the nineteen disasters: public and private employers, engineers, contractors, architects, trade unions, interest or pressure groups, local authorities, central government departments. Some of the organisations had gone out of business, changed name, or had been purchased. Telephone and letter contact was made with those organisations that could be located. Three standard letters were designed. One letter requested an interview for those organisations that were to be involved in the interview programme. The other two letters were designed for the organisations involved in the correspondence programme. Questions were formulated for the interview questionnaire based on knowledge of human factor failure. Twenty interviews, of which fifteen were taped with permission, were conducted in various informal and formal settings.

3.4 Data analysis

Individual abstract concepts were first of all derived from the raw data, after which a second analysis was performed using the same techniques to cluster these individual concepts into groups. This clustering into concept group types simplified data handling, and as a consequence made general patterns of organisational learning, where they existed, more readily identifiable.

The concept groups generated from the interview programme were then placed in two broader groups. First, a broad group of concepts of general applicability with relevance to many if not all organisational activities. Second, an activity-specific group of concepts relating to only one or, at the most, a small number of substantive organisational activities such as personnel management or the handling of organisational information.

It should be noted that there are no simple causal relationships between the concepts or their groups. Relationships which do exist are complex in structure, time and space. The view of relationships presented to the reader is not as rich as that which exists in the 'real world', because it is impossible to handle all these complex relationships simultaneously. As a result this exposition is one of many that could be selected from the evidence.

Notes

1 Brown, J. and Sime, J.A. (1981) *A Methodology for accounts in social method and social life*, Brenner, M.(ed.), Academic Press, London.

2 Davidson, D. (1973) In Defense of Convention T, in Leblanc, H. (ed.), *Truth, syntax and modality*, Proceedings of the Temple University Conference on Alternative Semantics, p 324, Amsterdam, North Holland.

3 Goode, W.J. and Hatt, P.K. (1952) *Methods in social research*, McGraw Hill, New York.

4 Webb, E.J. et al. (1966) *Unobtrusive measures = Nonreactive research in the social sciences*, Rand McNally, Chicago.

5 Arber, S. (1981) *Introduction to triangulation*. Unpublished manuscript, Sussex University.

6 Martin, P.Y., Turner, B.A. (1986) Grounded Theory and Organizational Research, *The Journal of Applied Behavioural Science*, Vol. 22, No. 2, pp 141-157.

7 Turner, B.A. (1989) *The way of the thesis*, Capriccio Press, London.

8 Marshall, C. (1985) Appropriate criteria of trustworthiness and goodness for qualitative research on education organisations, *Quality and Quantity*, 19:353-373.

Chapter 4

Generation of hindsight

Introduction

This chapter discusses the use of public inquiries to draw out the lessons from disasters. People have lost their lives in these incidents, and all involved in the inquiry process owe it to those people to bring out as full a picture as possible of what occurred in order to recommend ways of preventing recurrence of similar incidents. Some former witnesses to public inquiries are critical of aspects of procedure, especially the quasi-legal and adversarial nature of such investigations. Some have stressed that not all relevant issues may be considered by inquiries, that recommendations may be incomplete in scope, and that there may be too much emphasis on blaming someone or something.

However, despite these shortcomings it is likely that public inquiries will remain as the most valuable source of information to help prevent recurrence of disasters. It is important that causes are determined, that relevant recommendations are made, and that these recommendations are implemented. This chapter describes an analysis of public inquiry recommendations, and gives evidence to suggest that they sometimes do not address all of the causal chains of events leading to a complex socio-technical failure. To help remedy this a graphical method of analysing inquiry information and reports is described. This schematic report analysis method is a valuable tool both to those analysing disaster reports, and to those conducting inquiries to ensure that their scope is complete.

4.1 Public inquiries into disasters

In Britain the public inquiry system is a method of carrying out investigations into matters, events and circumstances which give cause for concern to the public and government. This system has developed over the last hundred years or so in a rather piecemeal way in response to the needs of public administration. There appears to be no clear definition of what a public inquiry is or how it should carry out its duties once constituted. Each new inquiry generally bases its composition and procedural framework on precedents from earlier investigations and from aspects of the British legal system.

The inquiries which are set up to address public anxiety after a major disaster follow this pattern. This is explicitly stated in the report of the Coldharbour fire:

> ...we have adopted a procedure which has been used at many previous Inquiries in particular Aberfan, Hixon and Ronan Point inquiries.[1]

An inquiry is usually set up under the provisions of legislation appropriate to the incident, where such statutes exist. For example, a mining disaster would be investigated under the Mines and Quarries Act, and an accident involving an airliner under the Civil Aviation (Investigation of Accidents) Regulations. Where no such specific legislation exists inquiries may be constituted under the Health and Safety at Work Act 1974. There are certain unspecified circumstances which, if sufficiently grave, can lead to an investigation being held under the 1921 Tribunals of Inquiry (Evidence) Act. The public inquiry into the circumstances surrounding the loss of HM Submarine *Thetis* is one example of an investigation convened under this infrequently used provision,[2] as is the inquiry into the coal spoil tip disaster at Aberfan, South Wales in October 1966. Although there have been many changes in health and safety legislation over the last 30 years, particularly the passing of the 1974 Health and Safety at Work Act, the 1921 Tribunals Act remains applicable under certain conditions.

4.2 Public inquiry procedures

There are variations across industries in the precise nature of the formal proceedings adopted. However, under whatever legislation they are constituted, all public inquiries held in the aftermath of a disaster appear to have a common factor: the quasi-legal and adversarial nature of the formal public examination of evidence relating to the incident. Inquiries take much of their philosophy, procedures and ethics from the British legal system. Typically each of the parties called upon to participate in the inquiry is represented by legal counsel. Each witness is examined and subsequently cross-examined by lawyers on the evidence they have submitted for consideration.

4.3 Criticisms of public inquiries

The research for this book included interviews with people who had attended public inquiries as witnesses. Some were impressed by the overall character of the inquiries. Questioned about the efficacy of one such inquiry a respondent replied that, 'it was the best fire seminar I ever went to'. Another remarked that, 'it wasn't the recommendations that have had the lasting impression, it was the whole process'.

While public inquires do appear to be generally well received, they are also the subject of some less positive criticism from those who have attended as witnesses. Some have commented unfavourably upon the quasi-legal nature of public inquiries outlined above. Others questioned findings, as this reply by a respondent illustrates:

There are passages in the conclusions of the report which need to be questioned.

Some felt that relevant points had not been considered at all. Asked for his criticisms, one manager commented:

I think the main one we felt was not a disagreement with the findings but an omission from the inquiry itself.

Another former witness was more blunt:

I more than disagreed. I saw it my own way.

A more formal example of criticism is provided in the 'Report of the Investigation into the Cause of the 1978 Birmingham Smallpox Occurrence'. The Report's foreword, written by the then Secretary of State for Social Services Patrick Jenkin, notes that:

The University have stated that they regard the outcome of the Court hearing as establishing that they were in no way at fault and that much of the assessment in this Report is substantially incorrect.[3]

4.3.1 Political problems

One set of criticisms of public inquiries may be put under the heading of political expediency. Some witnesses believe that considerations other than the evidence are sometimes taken into account during deliberations. As one interviewee put it:

> Some truths are frightening either in their effect on public morale or on the public purse! This can bowdlerise the final report.

Another observed that:

> Politically speaking you don't want inquiries to make their factual findings and come up with recommendations which cannot [be implemented], or give political difficulties in [implementation] and to a large extent inquiries are set up so that they come to the right conclusions.

A third respondent suggested that:

> ...politicians need to cover themselves against criticism.

A fourth suggested:

> It must be true that a good inquiry will find out things which are so disturbing that they can't be published or so drastic in their implications for social or economic or technical policy that they can't be implemented because resources do not exist to implement them fast enough to reassure the worried public.

4.3.2 Terms of reference

The inability of public inquiry participants to get particular issues raised often appears to stem from the setting of too narrow terms of reference. As an editorial in the *Architects Journal* noted:

> Surprisingly, inadequate terms of reference were given by the Secretary of State for Social Services to the chairman K G Jupp and his committee investigating the fire at the Fairfield old people's home.[4]

One reason why some respondents hold and express the cynical views outlined above may be that they believe that certain crucial issues are deliberately avoided. One senior manager stated that:

> The Inquiry went deeply into some aspects of the disaster, and persistently ignored others.

Another said that:

> ...there were other things like the management and staffing...not only were they not inquired into but attempts by us to get the latter one inquired into were given a pretty firm brush off.

Commenting on the public inquiry into the Fairfield Home fire, an article in the *Architects Journal* pointed out that:

> There is likely to be more than one theory, and the one chosen should be accompanied by an explanation of why it is more probable than the alternatives. The committee did not follow this course, and excluded mention of a theory which to my mind [the writer of the article] is a better interpretation of the facts, and this could affect the conclusions.[5]

A senior manager, when discussing this issue during the research for this book, replied that:

> I think it was felt by the Government that the persons who were holding the inquiry had sufficiently wide terms of reference to deal with anything and they covered the whole field. There may be others who would think otherwise.

A view expressed by a number of the interviewees is reflected in another respondent's comment:

> The definition of the terms of reference for an inquiry as for any Royal Commission are absolutely critical and a hell of a lot hangs on them.

4.3.3 Blame and truth

As already noted one repeated criticism of public inquiries was of their adversarial, quasi-legal structure. According to some respondents this appeared to fix the implicit goal of pinning the blame for the incident on someone or something. The *Architects Journal* noted that in the Fairfield inquiry:

> ...the trend of the lawyer's questions seemed not aimed primarily at establishing facts and arriving at the truth, or at apportioning responsibilities, but immediately at trying to prove theories evoked before the inquiry commenced and seeing if the blame for the disaster could be laid against the building system or against any individuals.[6]

One respondent expressed the view that:

> I think we were all looking over our shoulders hoping we wouldn't be blamed for anything.

Another interviewee related that he had:

> ...read the report of that [the Summerland fire on the Isle of Man] with growing fury that everybody was blaming everybody.

Similarly, Carver notes that when he attended a high-level conference called to discuss the conduct of mine disaster inquiries:

> It was openly stated by one representative, and agreed to by all parties, that the first casualty in their mine disaster Inquiry was truth and that their Inquiries concentrated more on the search to blame rather than on the search for truth: on fault finding rather than fact finding.[7]

4.4 Are public inquiries appropriate?

Given the above criticisms it is hardly surprising that many of the respondents feel that public inquiries are inappropriate vehicles for the investigation of disasters. A manager in one large organisation stated that:

> ...there were things they did not want to know and that shattered my bloody faith in Public Inquiries.

Another interviewee commented that in his opinion:

> ...the quasi-legal style of inquiry is not altogether conducive to getting the truth out of people.

One respondent stated that he felt that:

> ...there's a lot wrong with the present procedures.

and yet another argued that:

> ...the adversarial process is not the way to getting to the root of the problem.

A summary of the objections was provided by a manager who commented that:

> It seems to me in having the adversarial system, the attitude of all the protagonists who are involved is to try and protect their own interests...they are not interested in anything that may lead to the truth.

To the above criticisms one could also add the fact that public inquiries have no power to ensure that their recommendations are implemented. Therefore, it could be argued that they are in some case little more than devices employed by governments to pacify the public. It seems that a strong argument can be made for the introduction of a different form of inquiry into disasters.

One way forward which meets some of the aforementioned criticisms is adoption of an informal inquiry system. This might further the pursuit of truth and help prevent people from hiding matters that would best serve society by being out in the open. In this form of investigation no lawyers would be present, and the proceedings would be conducted informally and without pomp and circumstance. Witnesses would be asked questions by those persons eminent within the particular sphere of technical or social expertise relevant to witnesses' evidence. The emphasis would be upon learning how to prevent the incident from recurring rather than upon who or what was to blame. To assist this end, some version of the Confidential Human Factors Incident Reporting Programme (CHIRP), the confidential reporting system used by the UK aviation industry could also be instituted more widely. Whatever the precise format, practically all of those interviewed during this research appeared to favour such a less legalistic form of inquiry procedure.

Regardless of the shortcomings outlined above public inquiries are likely to remain as the best source of information regarding the causes of disasters and the means of preventing recurrence of similar events. These twin goals are described explicitly by Lord Cullen in the introduction to the report into the Piper Alpha disaster:

> Through the Inquiry I sought the answers to 2 questions.
> – What were the causes and circumstances of the disaster on the Piper Alpha platform in July 1988 and
> – What should be recommended with a view to the preservation of life and the avoidance of similar accidents in the future?[8]

Despite criticisms, it is important that public inquiries are fully exploited to draw out the lessons of disaster causation and prevention. Causes must be determined, recommendations must be made and then implemented. The rest of this chapter looks at how this process may be carried out more completely and efficiently.

4.5 Analysis of recommendations

The research underlying this book included analysis of the recommendations of nineteen public inquiries to see if any common learning patterns would emerge. The analysis revealed that individual inquiry recommendations covered some common ground. Various types of recommendation were coded and a written specification defining each type was drawn up. After examining all the recommendations from the inquiries under study, it was found that they could be placed into either a technical or a social category. These two major categories were then subdivided into five distinct groups, and these groups in turn further subdivided into 24 separate recommendation types. Three recommendations from the case studies did not fit into any of the type subdivisions, but since they only appeared once each they were not classified as a 'type', no definition was derived for them, and they were left in a residual category.

The five groups of recommendations identified were as follows:
  (A) Technical;
  (B) Authority-based;
  (C) Information;
  (D) Personnel;
  (E) Attempted foresight.

With their respective recommendation types they are illustrated graphically in Figure 4.1.

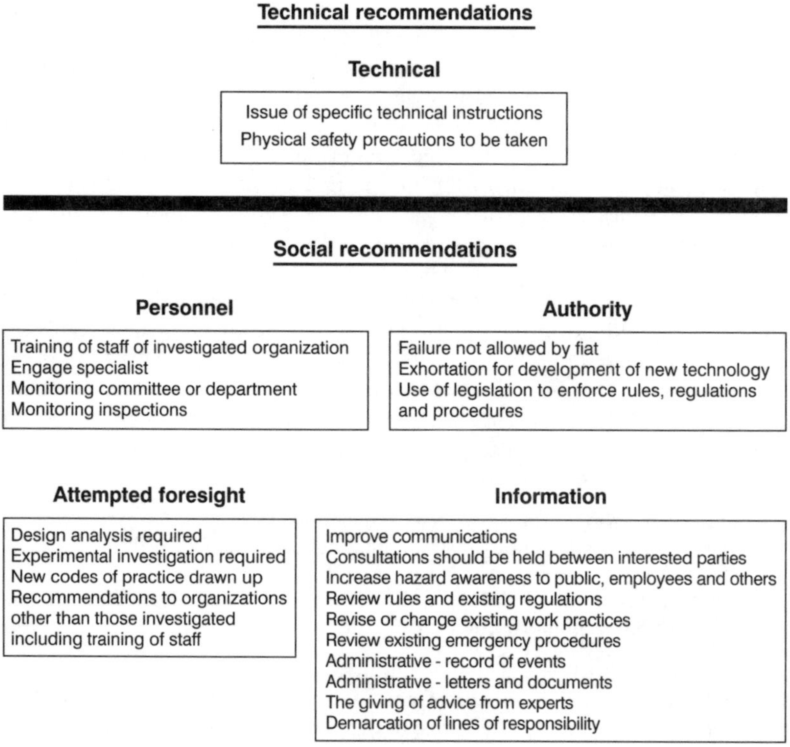

Technical recommendations

Technical

Issue of specific technical instructions Physical safety precautions to be taken

Social recommendations

Personnel

Training of staff of investigated organization Engage specialist Monitoring committee or department Monitoring inspections

Authority

Failure not allowed by fiat Exhortation for development of new technology Use of legislation to enforce rules, regulations and procedures

Attempted foresight

Design analysis required Experimental investigation required New codes of practice drawn up Recommendations to organizations other than those investigated including training of staff

Information

Improve communications Consultations should be held between interested parties Increase hazard awareness to public, employees and others Review rules and existing regulations Revise or change existing work practices Review existing emergency procedures Administrative - record of events Administrative - letters and documents The giving of advice from experts Demarcation of lines of responsibility

Figure 4.1 Conceptual grouping of public inquiry recommendations

4.5.1 Aims of recommendation types

The groups of recommendations identified are respectively intended to achieve the following aims.

(A) Technically based recommendations attempt to prevent the same or similar incidents from taking place by recommending that currently manufactured physical safety precautions be installed where they appear to be required. This group consists of two recommendation types:

1. *Safety precautions to be taken:* here it is stated that specific physical or procedural safety measures should be adopted in addition to those which were in force at the time of the incident.
2. *Issue of specific technical instructions:* recommends precise changes to the organisation's everyday working practices, procedures, rules, materials or artefacts.

(B) Authority based recommendations appear to be made in an attempt to produce safety by demanding it. They include recommendations that new legislation, rules, orders, etc. be introduced to prevent people or organisations from performing certain actions; or they may exhort the development of a new technology which could prevent another similar disaster or ameliorate some of the consequences. This group consists of three recommendation types:

1. *Failure not allowed by decree:* a recommendation which effectively orders that further disasters similar to that which has occurred must not occur again in the future. An example is a statement such as: 'The possibility of an explosion must be recognised when designing'.
2. *Use of legal powers to ensure new or old regulations are adhered to:* this recommends that the full weight of the law be used to deter organisations or individuals from ignoring existing or new regulations.
3. *Exhortation for the development of a particular technology:* here it is held that a particular nascent technology should be developed as quickly as possible.

(C) Information based recommendations are designed to improve the communication of information between individuals, departments and other organisations and in some cases the wider general public. They are also aimed at assisting the organisation investigated to clarify ambiguous situations, by for example, recommending new administrative procedures to be adopted so that the demarcation of lines of responsibility between departments or organisations are clearly defined. This group of recommendations also aims to prevent future failures by the use of *a posteriori* precautions.

These include recommending changes in existing rules and regulations, and encouraging regular revision of existing work practices and procedures to ensure that they remain appropriate for the task in hand. This group consists of ten recommendation types:

1. *Review current rules and regulations:* it is held that the current rules and regulations relevant to the incident be reviewed in the light of the occurrence, with changes to be made if that is deemed to be appropriate.
2. *Revision or change in existing work practices and procedures:* this aims to ensure that organisations revise work practices or procedures in the light of the incident which has occurred.
3. *Administrative, records of events, manuals, etc:* some form of administrative action is recommended, either: (a) the production of manuals to assist the memory of an organisation's staff or to aid the following of specified procedures; or (b) requiring staff to keep records of events that have taken place; or (c) requiring that staff should be issued with or have to complete certificates before being allowed to proceed with certain aspects of their work. An example of the latter would be a permit to use a welding torch in an area where it would not normally be allowed, say a fuel farm.
4. *Improve communications:* a recommendation that there should be either a physical arrangement (for example, loudspeakers installed) or some other procedure instituted which will assist more effective communication of relevant information between those participants involved in the work processes associated with the accident's occurrence.

5. *Consultations should be held between interested parties:* in this case it is stated that consultations should be held between all the parties involved in a project. These parties would typically include the relevant public authorities: for instance, the local town council planning office or fire brigade. The intention behind such a recommendation is usually that there should be full and free exchange of information between the various parties, so that all concerned have the clearest possible picture of what is to take place before the project design is finalised.

6. *Increase awareness of hazards to public, staff or other parties:* it is held that anybody who may come into contact with a known hazard should be made as aware as possible of the danger. Typically, recommendations call for use of notices, signs and posters.

7. *Review existing emergency procedures and practices:* a recommendation that emergency practices and procedures of an organisation should be reviewed as a result of shortcomings discovered.

8. *Demarcation of lines of responsibility to be drawn up:* it is stated that there should be an explicit organisational hierarchy of control, so that employees of an organisation know exactly who is responsible for what duties and to whom they should report within the enterprise.

9. *On the giving of advice by experts:* observations are made about the status and nature of 'expert' advice. This involves advice that was obtained before the incident or that which should have been sought but was not.

10. *Administrative letters, documents, signs, pamphlets to make existing rules and regulations clearer:* here it is noted that written material to aid in the interpretation of existing rules or regulations is, or could be, advantageous in reducing the possibility of misrepresentation.

(D) Personnel based recommendations are designed to bring personnel problems to the attention of management, such as lack of training of staff or the advisability of calling in specialist help rather than trying to manage with in-house knowledge. Also such recommendations include those designed to increase the amount of supervision exercised over individuals, departments, organisations or groups of organisations through the use of inspectors or monitoring committees. There are four recommendation types in this group:

1. *Monitoring inspections to be carried out:* a recommendation that both the work practices and procedures used in a current working situation should be inspected regularly. The purpose is to ensure that: (a) practices and procedures are carried out correctly – that is, people employed to carry out particular tasks are complying with existing rules; and (b) the practices and procedures are still relevant to the current work environment.

2. *Training of staff of investigated parties:* it is held that some or all of the staff in one or more of the bodies involved in the investigation should receive some formal training, which it is believed will assist in preventing a recurrence of the incident.

3. *Committee or department to monitor matters of safety:* a recommendation that a committee or department should be set up specifically to continually monitor all aspects of safety within the investigated organisation or similar.

4. *Engage a specialist:* it is stated that a full or part-time specialist should be appointed to monitor aspects of safety of the organisation concerned, with particular emphasis upon that part of the organisation which was involved in the disaster.

(E) Attempted foresight recommendations are generated by an inquiry in an attempt to forestall problems which could arise in the future. This is done, for example, by making recommendations to organisations other than the ones involved in the inquiry, or by suggesting that experimental investigations be carried out in order to advance knowledge in particular fields so that information will be available to assist future designers and managers of socio-technical organisations. There are five recommendation types involved in this group:

1. *Experimental investigations required:* a recommendation that experimental investigations, either in the laboratory or on site, should be carried out: (a) to determine more accurately the nature of some aspect of the accident; (b) to ensure that in future, by carrying out on site tests, a disaster similar to the one which has already taken place will not recur; (c) to advance the knowledge of particular technologies involved with the incident, so that the information generated will be available to give advance warnings in future about possible recurrences of that particular disaster scenario.

2. *Design analysis required:* here it is held that some aspect of the accident which has come to light as a result of the investigation either should have had, or should have, a design analysis conducted upon it so that the particular artefact in question will be less likely to contribute towards a failure in the future.

3. *New codes of practice to be drawn up:* the Board of Inquiry directs that there should either be a change in the present code of practice to accommodate the findings of the Board, or that a code of practice should be drawn up if no current code exists.

4. *Recommendations to bodies other than the ones investigated:* here recommendations are made to particular parties other than those directly connected with the disaster under investigation – for instance, the professional institutions, organisations such as the Royal Institute of British Architects, or the Fire Protection Association.

5. *Training of staff other than the parties investigated:* a specific reference to the training of staff in organisations not directly involved with the disaster under investigation. For instance, the training of architects, police officers or engineers, whose professional skill in the opinion of the Board would be enhanced by incorporating the findings of the inquiry into their teaching material.

4.5.2 Groups of recommendation types

Analysis of the nineteen case studies showed a distribution of recommendation group types as follows:

Recommendation group type	Disasters with type (%)
Information	100
Technical	100
Attempted foresight	85
Personnel	85
Authority	60

Three possible corollaries can be drawn from the fact that virtually all the recommendations analysed can be located in one of these larger group types. First, that investigations into accidents only concentrate on a number of specific areas. Second, that there are only a limited number of ways in which organisations can learn from their mistakes. Third, that the underlying causes of accidents may be similar in nature. If this clustering of recommendation types occurs because investigating bodies only concentrated on particular areas, then some learning opportunities might well have been lost.

However, if clustering occurs because organisations are limited in the number of ways in which they can learn then it is of paramount importance that we gain as much knowledge of these areas as possible in order to extract the maximum amount of learning when such situations are presented to us.

Alternatively, if the clustering is due to accidents having similar aetiologies, this implies that it might be possible to reduce the risk of similar incidents occurring within a wide range of organisations if predisposing mechanisms can be identified.

As noted earlier public inquiry recommendations are intended to prevent recurrence of similar incidents by effectively breaking the chains of unwanted causal events and conditions.

Recommendations are derived from investigation of the causes and circumstances of a disaster, and they mirror the nature of these causes and circumstances. Hence, analysis of recommendations not only indicates what kind of learning has taken place but also the organisational problem areas that led to the incident. From this analysis it can be seen that some of these problem areas are:

1. selection of appropriate physical safety precautions;
2. supervision and monitoring of processes, procedures, individuals, departments and organisations;
3. keeping working practices, procedures, rules and regulations up to date;
4. exchange and management of information between individuals, departments and organisations;
5. training of staff and the updating of their skills;
6. identification of ambiguous situations;
7. lack of aptitude in using what Pidgeon has termed 'safety imagination'. This is embedded within the concept of safety culture, and is the ability of management to imaginatively reflect upon their current or past situation and, in doing so, to foresee some of the problems which might arise and implement policies to ensure that they do not occur.

As discussed in Chapter 1 , the initial reporting of a socio-technical disaster typically emphasises a technical fault as the major reason for the disaster occurring. However, one interesting observation to be drawn from the above analysis of public inquiry recommendations is that the majority are concerned with the problems of management, administration and information rather than technical matters.

This is not to argue that technical causes of disasters are of little or no importance, but that they constitute only one aspect of a multi-faceted picture. Accordingly, many of the ultimate recommendations from disaster investigation centre around 'softer' problems of management rather than 'hard' technical ones. This reflects the socio-technical nature of complex failures.

An analysis of public inquiry recommendations, as Turner and Toft note:

> ...displays to us both the areas which were of concern to those conducting these inquiries and the model of diagnosis and prevention which the inquiry tacitly adopted. These particular public inquiries sought to control hazards and to prevent the recurrence of major incidents by advocating action at a physical, administrative and at a communications level as well as sometimes proposing actions which ranged more widely where future plans to deal with a particular hazard were concerned. The model behind these arrays of practical recommendations stresses the importance of: selecting appropriate physical safety precautions; identifying and eliminating ambiguous situations; keeping working practices, rules and procedures up to date; training staff appropriately; improving communication about hazardous matters; and attending to the supervision and monitoring of processes and individuals within the organisation concerned.[9]

4.6 Schematic report analysis

One of the central problems in attempting to learn from large-scale accidents is that of how to marshal the evidence so that the relationship between the interacting networks of causal events can be observed. Only if this process is done efficiently can recommendations be drawn to prevent recurrence of an incident.

The technique of schematic report analysis was developed by Turner[10] to highlight how disasters do not spring into existence immediately and in fully developed condition. Rather, major accidents take form during an 'incubation period' when there is an accumulation of undetected sets of events, which may lie dormant until one event triggers the disaster. The schematic report analysis technique has been used to summarise a number of public inquiry reports[11] as well as other kinds of unforeseen incidents such as the Yom Kippur War,[12] and to analyse cases of structural failure.[13]

Schematic report analysis aids the examination of public inquiry reports. Typically the evidence in a public inquiry is taken from witnesses, both 'expert' and 'lay', and is then synthesised into a formal written report giving an authoritative account of the incident. This synthesis gives the answers to the question: 'What were the causes and circumstances of the disaster?' It is normally accompanied by a set of conclusions and recommendations for action. Generally such a procedure works efficiently, but there are grounds for arguing that some opportunities for learning may be lost in this process. The comments noted above from some public inquiry attendees suggest that this is the case. In part this is due to the complex nature of socio-technical failures.

The written report of a public inquiry often runs into many thousands of words so that it is sometimes difficult for the interelated sets of events to be fully appreciated. It may then be difficult to extract all the lessons from an incident so that they can be made available to those who need them. In some complex accidents, it may prove difficult for members of the investigating team to comprehend fully all the implications presented by the evidence at their disposal, with the result that they may unwittingly present a limited set of recommendations. 'Blindspots' may have been created by the complexity of the incident. While a great deal of learning does take place from public inquiries with their present techniques, some relevant lessons may not be discovered.

4.6.1 Properties of a schematic report analysis diagram

It is possible by using the schematic report analysis technique to translate the written synthesis of a public inquiry report into a 'simpler' visual presentation of the different chains of events built up during an incident's incubation period, as determined by the inquiry. Such a visual presentation may assist the members of the investigating team to clarify their own diagnosis of events, and use of the technique is worth considering on these grounds alone. However, it is also possible to use such a diagram to link the diagnosis of the incident to the recommendations, and this can reveal valuable information as will be discussed below.

The construction of a schematic report analysis diagram (SRAD) necessarily involves some simplification and typification of events, as does the drafting of the original accident report. However, the advantage of being able to summarise large amounts of information in a readily comprehensible form more than compensates for any simplifying assumptions which have to be made in the process of translation. Figure 4.2 sets out the diagram for a mining accident, showing how it is possible to summarise a lengthy report in a single presentation. The convention adopted in this diagram has been to enclose in solid boxes those conditions fully appreciated before the incident, and in broken line boxes those conditions which were hidden or only partially understood before the events. This convention need not be used in all cases.

Figure 4.3 illustrates how it is possible to relate recommendations to the model of event chains. In the case of the incident described it can be seen from the model how a learning opportunity was in fact lost to the investigating committee: three chains of contributing events were not dealt with by any of the recommendations of the inquiry.

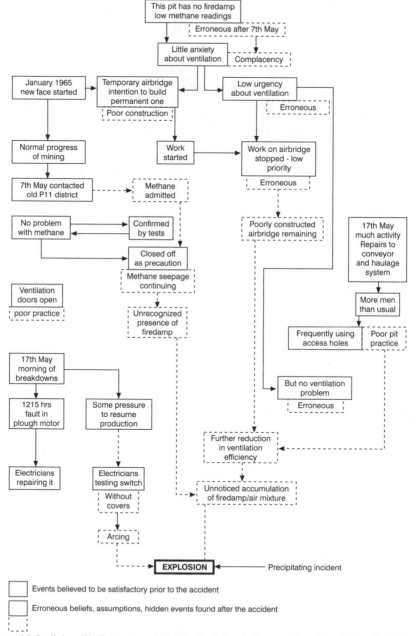

Figure 4.2 Outlining of individual event SRAD. Cambrian Colliery accident (from Turner[10], 1978)

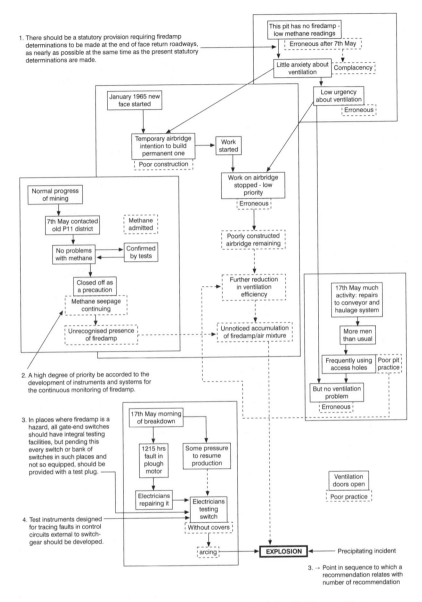

Figure 4.3 SRAD showing main clusters of events for Cambrian Colliery accident

4.6.2 Preparation of a schematic report analysis diagram

The information input to the form of SRAD described above is garnered from the final draft report of a public inquiry or similar investigation. By using earlier drafts to generate an SRAD it would be possible to include the diagram so produced in the body of the final report. This iterative process would help to guide the Board of Inquiry in its investigation and to disseminate its findings in a more comprehensible form.

In preparing an SRAD the following steps are taken. From the report's account of activities and circumstances leading up to the incident in question, a note is made of all events, from any source, which the inquiry team finds to have had a bearing on the development and production of the incident. Using the information extracted from the text of the inquiry report, chains of such events are assembled in chronological order, and in a way which also illustrates those causal links which have been determined or proposed by the inquiry.

When these chains of events have been established, links between events and chains can then be sought, and the causal network leading up to the incident roughly modelled. This makes it possible to indicate, for instance, where actors have made false assumptions, held erroneous beliefs or have been in the habit of using poor working practices. The purpose of making such indications on the diagram is to try to ensure that the more general types of dysfunctional organisational behaviour are highlighted so that appropriate remedial action can be recommended. For example, organisational personnel may have drifted into some form of complacent behaviour which assisted in the development of the incident. This complacency may have manifested itself in the spread of erroneous beliefs about their working environment, which in turn had some bearing on the incident which occurred. Highlighting these shortcomings in an organisation's safety culture could then bring to the investigators' attention the need for general recommendations relating to training standards and motivation.

Typically in the early stages the diagram will be very complex and disorderly. The disorder may be dealt with by reducing to a minimum the amount of detail in the model insofar as this can be achieved without losing those events which are believed to be of significance, in the judgement of the compiler. When this process has been taken as far as seems reasonable, the whole model can then be drawn up graphically to provide a schematic diagram of the interlinking chains of events leading up to the incident.

The resulting diagram will serve to communicate both to the accident analysts and the wider public the complex events leading up to the incident in question. It does this by means of a single visual presentation, albeit in some cases one of considerable complexity. The diagram also serves as a base to relate any recommendations which are forthcoming. Accordingly, it also indicates those contributory chains of events which are not dealt with in the report's recommendations, prompting the investigators to provide a rationale for such omissions.

4.6.3 Specific examples of SRAD

The whole process can be more readily understood through the consideration of specific examples. A firedamp explosion took place at the Cambrian Colliery, Glamorgan, on the afternoon of 17 May 1965. Of the events leading up to the disaster perhaps the most important factor in its creation was the belief that a seam was gas free. In the words of the official report:

> Everyone concerned at the colliery regarded the Pentre seam and P26 face in particular as virtually gas free. The management were no doubt fortified in this belief by the fact that the statutory mine samples never showed more than 0.3 per cent firedamp.[14]

The colliery was therefore known by those who worked there to be one in which only small amounts of firedamp were present, a condition continually reaffirmed through regular checks. It followed that there was thought to be little need to worry about ventilation and procedures associated with it. The events shown in Figure 4.2 are as they came to be known after the explosion had taken place. The following résumé of the incubation period related to the disaster is taken from Turner:

> Reviewing the events presented in Figure 4.2, it can be seen that in this 'virtually gas free' pit, a new face came into production in January 1965. In the process, a temporary air-crossing between two tunnels had been constructed with access holes, rather than a full air lock, to allow the men to pass from one tunnel to the other without disturbing the ventilation across the face. The construction of this air-crossing was not of very high quality, but ventilation was not a high priority issue: in any case it was intended shortly to build a permanent air-crossing, and work had started on this, although it had been suspended, because of the low urgency of ventilation, to allow other more important work to go ahead.
>
> In the course of the normal process of mining, an old district (P11) had been contacted on the 7th May. There was not thought to be any problem here with methane, and this was confirmed by tests, but the link had to be sealed off anyway, and after six days' work this was completed, methane tests continuing to be negative. On the 17th May, a pair of ventilation doors linking another face were thought to have been left open. There was much repair activity on the morning of 17th May, owing to breakdown and repairs to the conveyor and the haulage system. This brought more men than usual to the P26 face, and required them frequently to use the access holes in the temporary air-crossing, a matter which was not particularly remarkable in the absence of ventilation problems. Following the breakdowns in the morning, there was some pressure for production to resume, when at 1215pm there was an electrical fault in the plough motor. Electricians were called to repair it and at 1240pm they were testing the switch in connection with this repair. The testing of this switch, which did not have its gas-tight cover correctly bolted on, ignited the firedamp-air mixture.[15]

In Figure 4.3, the various branches of the causal analysis have been isolated, and the recommendations of the inquiry are shown related to the events which they are designed to prevent recurring. As can readily be observed the schematic diagram can be broken down into six separate yet interelated event chains. Yet, the recommendations made as a result of the inquiry only seek to intervene in three of those chains.

Recommendation 1 would appear to be concerned to prevent the firedamp chain of events from recurring by trying to ensure that tests for methane are carried out in return headings with the same frequency as they are elsewhere in the mine. Recommendation 2 appears to be aimed at preventing firedamp from being able to accumulate unnoticed by mine personnel. Recommendations 3 and 4 seem to be designed to prevent mining staff from exposing circuits prone to arcing to the atmosphere while tests are being carried out.

Thus, the recommendations attempt to stop a recurrence of the disaster first by trying to use both regulations and new technological practices to cut out any chance of an unnoticed build up of firedamp. Secondly, they recommend the introduction of other technical changes by encouraging the development of test equipment to prevent technicians having to open mining equipment while underground in order to carry out fault diagnosis. Together the two kinds of remedial action recommended by the Board of Inquiry try to prevent the physical process of a spark igniting an unrecognised build-up of firedamp.

What the recommendations do not do, as an inspection of the causal schematic model shows, is deal with the more fundamental problem which underlay the accident, that of poor mining practice. In some ways this is rather surprising since the conclusions to the report state that:

> Much of the Inquiry turned upon the air bridge over the Mindy Heading Return...the poor construction of the air bridge and the way in which it was used was deplorably bad pit practice, which should not have been tolerated by those having responsibilities for the safe working of the mine.[16]

Similarly, the report notes that the process which the electricians were engaged in at the time of the disaster was known to be a boring one, and that:

> ...one can, therefore, understand but not excuse, electricians who are tempted to avoid this tedious process...it is imperative that they recognise the danger and do not fault in this respect.[17]

However, no recommendations are made in the report to deal with these rather severe shortcomings of practice and behaviour to prevent their recurrence. This may have been due to an oversight on the part of the investigators, or by their inability to formulate an appropriate course of action due to the cognitive load created by the wealth of evidence presented to the inquiry. Such an omission, it could be argued, would be less likely to occur if the investigators of an accident were to utilise the SRAD technique. Figure 4.3 highlights how the relationship between an inquiry and its recommendations can be more clearly and immediately seen than would have been the case in a solely textual report.

A second example to illustrate the application of an SRAD may be drawn from the report on the collision between a Manchester-Euston express train and a road transporter carrying a 120 ton electrical transformer at the automatically operated half-barrier crossing at Hixon, Shropshire on 6 January 1968.[18]

Unlike the Cambrian Colliery explosion, in this accident several different organisations were involved, each playing a role in the creation of the incident. The task of unravelling the underlying causes of the incident and producing effective recommendations to prevent a recurrence of the event was correspondingly more difficult. Evidence was sought within British Rail, from the train crew, from those departments responsible for planning and implementing the introduction of the new form of automatic crossings, from those who were responsible for disseminating publicity to potential users of the crossings, and from those who installed, inspected and modified the crossing at Hixon.

From within the Ministry of Transport, evidence was taken from the Railway Inspectorate responsible for approving the new crossings and the procedures associated with them. From within the police force, those responsible for circulating information about the automatic crossings, and those responsible for briefing the police patrols escorting abnormal loads were questioned, as well as the officers who were escorting the load involved in the collision. The road haulage contractors were a further group questioned.

An inspection of Figure 4.4, which is set out in schematic form, clearly shows how in such a case the complex relations between different chains of events involving several organisations can be mapped to produce an overall model outlining the central features of the incident. In turn this could be of assistance to the investigating team when considering their recommendations.

In this example of the SRAD technique the intention is not to dwell upon the actual events of the incident which are fully described in the accident report.[19] Here we use the accident to further illustrate the way in which an SRAD can be produced, and to indicate how different levels of generality can be dealt with. Therefore, for ease of presentation the text of the individual event boxes has been omitted. Figure 4.4 shows how the individual events chosen as being of importance in the production of the incident can first be used to construct the detailed SRAD model. Then, by considering each chain of events as a single entity, it is possible to produce an SRAD model at a higher level of generality, as in Figure 4.5. Finally, by clustering the single chains of events together one can produce another SRAD model of even higher generality, as in Figure 4.6. The level of generality at which SRADs of the incident are eventually presented, is of course, at the discretion of the investigators.

Inspection of Figure 4.5 shows how the different individual events shown in Figure 4.4 have been clustered together to become single chains of events so that it is a far less complex

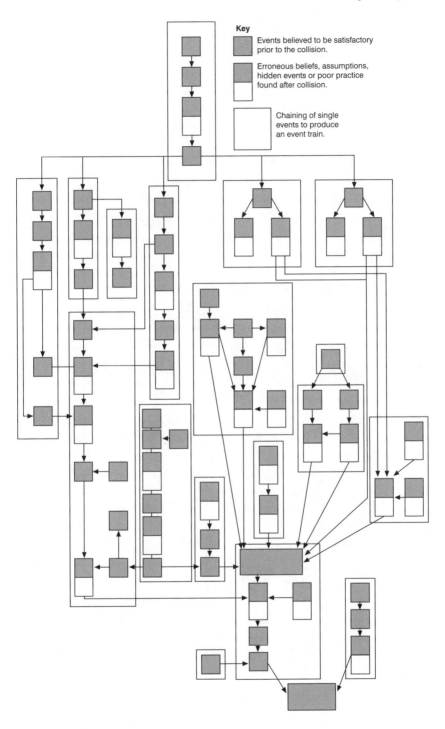

Figure 4.4 Outline of individual-event SRAD. Hixon level crossing accident.

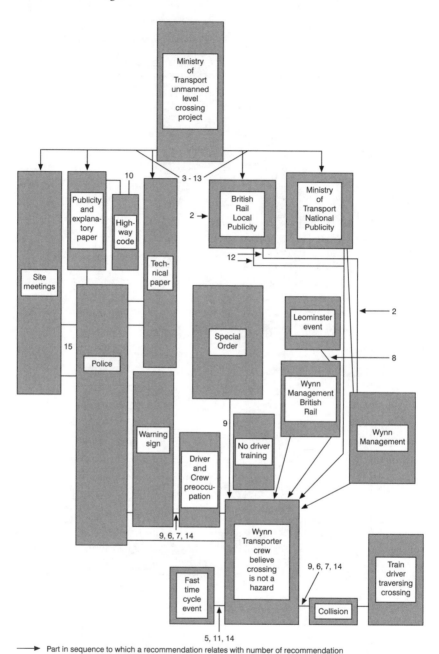

Figure 4.5 SRAD showing the main clusters of events for the Hixon level crossing accident

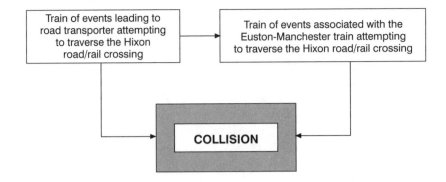

Figure 4.6 Highest level of SRAD analysis for Hixon level crossing accident

diagram than its predecessor. It does, however, still show all the sets of events that are believed to be of central importance in the creation of the incident. The event chains are represented by single boxes and identified by a title which reflects their group nature, for example 'Police' or 'British Rail Local Publicity'.

Figure 4.5 highlights both the relationship between the various event chains, and between the public inquiry recommendations and the event chains. The recommendations can be thought of as attempts to sever the linkages of the individual chains, in the hope of preventing recurrences of similar incidents. The text of recommendations intended to block off particular chains may be displayed on the diagram, as in Figure 4.3, or be referenced to the appropriate point in the report, as in Figure 4.5.

A further inspection of the SRAD in Figure 4.5 reveals again that some event chains do not have recommendations tied to them. For example 'No Driver Training', 'Wynn Management British Rail' or 'Technical Paper'. Again, it is possible that the potential for learning from a serious accident may not have been fully exploited.

4.6.4 Computer application of SRAD

The form of SRADs lends itself ideally to representation and analysis on computer 'windowing' systems. Once an SRAD has been generated by a human compiler it is possible to store the sets of related diagrams as nested sets using proprietary software programs such as Macintosh Filevision. VDU displays at different levels of generality can then be generated. An example, using the Cambrian Colliery explosion discussed above, illustrates the concept in Figures 4.7 – 4.10.

The Filevision program allows a diagram to be drawn using a standardised set of computer graphics. Once the diagram for a particular level of SRAD generality has been created, as in Figure 4.7 for example, it is possible for the user to identify to the computer each of the chain of event boxes that compose the diagram. Therefore, when an SRAD of a different level of complexity has been generated, as in Figure 4.8, it is possible via the software to link the two SRADs together.

When a user wishes to see in more detail the chain of the events which created the condition indicated by a particular box, this can be done by selecting that box using a 'windowing' technique. This means that when a box is selected for investigation, the next lower level SRAD contained within that box is presented for viewing, like looking through a

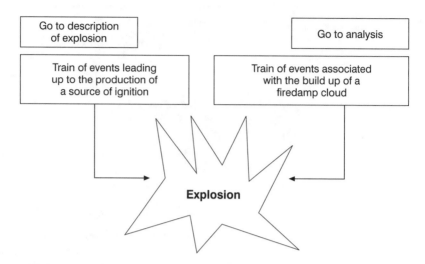

Figure 4.7 Schematic report analysis diagram. Cambrian Colliery explosion, level 1

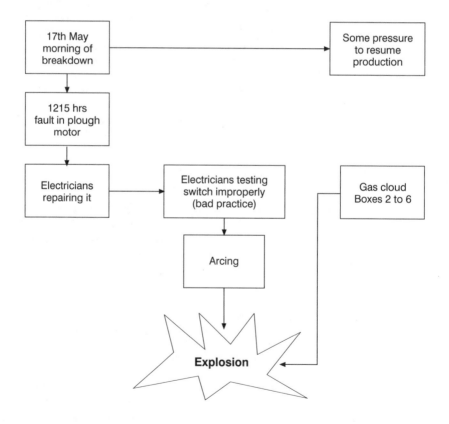

Figure 4.8 Schematic report analysis diagram. Cambrian Colliery explosion, level 3

Events associated with firedamp build up BOXES 2 - 6
BOX 2
old district
events train
BOX 3
airbridge
events train
BOX 4
ventilation
events train
go to recommendations
go to analysis
go to description
BOX 5
haulage and
conveyor
events train
BOX 1
source of
ignition
Explosion
BOX 6
poor practice

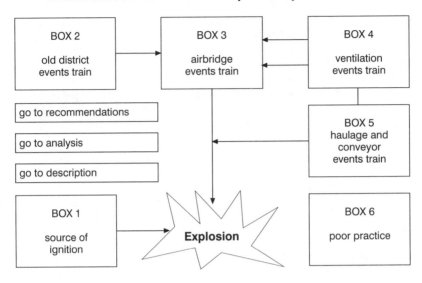

Figure 4.9 Schematic report analysis diagram. Cambrian Colliery explosion, level 2

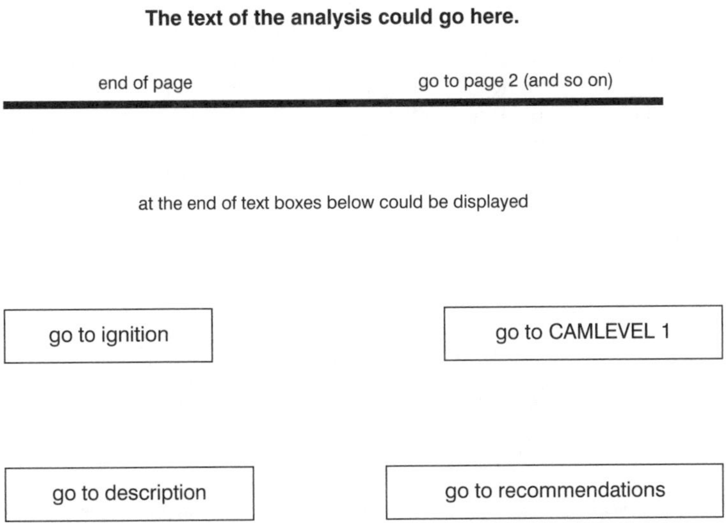

The text of the analysis could go here.
end of page
go to page 2 (and so on)
at the end of text boxes below could be displayed
go to ignition
go to CAMLEVEL 1
go to description
go to recommendations

Figure 4.10 Schematic report analysis diagram. Cambrian Colliery explosion analysis

window. For example, selection of the box 'Trains of Events leading up to the Production of a Source of Ignition', in Figure 4.7, would reveal the SRAD in Figure 4.8. Similarly if the box 'Gas Cloud' was selected in Figure 4.8 the SRAD in Figure 4.9 would be revealed.

Using this technique it is also possible to put additional reference boxes in the diagram. For instance, windowing on the 'Go to Analysis' box in Figure 4.7 would present the user with the relevant textual analysis of the disaster from the inquiry report (see Figure 4.10). Having read the analysis the user could then select the next aspect of the incident of interest.

If this approach were to be used in a structured teaching environment for training staff concerned with disaster management, it would be possible simply to guide the students through the incident in a particular way, with set questions for them to answer as part of their written course work at the end of the presentation.

4.7 Conclusions

Public inquiries offer an institutionalised means of learning from catastrophic events. While they can be criticised in some respects, public inquiries do provide through their formal standard structure and procedures a recognised opportunity for analysis of disasters. The analysis may subsequently be used to isolate the crucial events which brought about the disaster and to offer remedies to prevent similar events recurring. However, it is clear that many professional people, especially those who have acted as witnesses to inquiries, would like to see changes in the way they are carried out. In particular they suggest a move away from a legalistic format.

It is also clear from the foregoing discussion that disasters are the result of a series of complex interacting events. Even though the complexity of some incidents may make it impossible to achieve a total understanding of all that has taken place, useful analyses can nevertheless be carried out. Such analyses show quite clearly that while the technical aspects of any disaster are important, those elements which relate to management, information and personnel are equally, if not more so.

Because of the complexity of socio-technical disasters, it is often extremely difficult for an investigating team to appreciate all the evidence that is presented to them. However, it is of the utmost importance that the maximum amount of information is derived. The schematic report analysis diagram (SRAD) technique outlined above could be of help in pursuing this goal. The technique may not ensure that all lessons from a given incident are uncovered. But in both its computerised and manual forms it does provide a further tool to be added to those which already exist for the investigation and analysis of unforeseen incidents in socio-technical systems.

Notes

1 *Report of the Committee of Inquiry into the fire at Coldharbour Hospital, Sherborne on 5th July 1972*, Cmnd 5170, HMSO, London.

2 *Royal Commission on Tribunals of Inquiry*, (1966) Cmnd 3121, p 53.

3 *Report of the Investigation into the cause of the 1978 Birmingham smallpox occurrence* (1980) HC668, HMSO, London.

4 *Architects Journal* (1975) 10 October.

5 *Architects Journal* (1975) 16 July.

6   ibid.

7   Carver, J. (1981) 'The options available for mine disaster inquiries' *Ignitions, explosions and fires.* Procs. Symp. - University of Wollongong, 12 - 15 May.

8   *The Public Inquiry into the Piper Alpha Disaster*, (1990) Cm1310, November, London, HMSO.

9   Turner, B.A. and Toft, B. (1988) Organisational learning from disasters, in Gow, H.B.F. and Kay, R.W. (eds.), *Emergency planning for industrial hazards*, Elsevier Applied Science.

10   Turner, B.A. (1978) Man-made disasters, Wykeham Publications, London.

11   Toft, B. (1986) *Schematic report analysis diagramming: an aid to organisational learning*, unpublished manuscript, Exeter University.

12   Stech, F.J. (1979) *Political and military intention estimation*, Mathtech Inc., Bethseda, Maryland.

13   Pidgeon, N.F., Blockley, D.I. and Turner, B.A. (1986) Design practice and snow loading – lessons from a roof collapse, *The Structural Engineer*, Vol. 64A, No 3, March.

14   *Explosions at Cambrian Colliery, Glamorgan* (1965) Cmnd 2813, London, HMSO.

15   Turner, op. cit.

16   *Explosions at Cambrian Colliery*, op. cit.

17   ibid.

18   *Report of the Public Inquiry into the accident at Hixon level crossing on 6th January 1968*, (1968) Cmnd 3706, HMSO, London.

19   ibid.

Chapter 5

General organisational learning

Introduction

This chapter draws heavily on the responses from those interviewed for the research programme underlying this book. It looks at these responses in terms of eight general concept groups derived from analysis of the research: organisational isomorphism; emotional impact; chance and disaster; attempted foresight; hindsight; organisational reaction; safety by compulsion; and safety philosophy. These concept groups may be viewed as general factors that can influence the ability of organisations to take steps towards 'active foresight'. Active foresight is the goal of the organisational learning process: it combines foresight of the possible causes of disaster, with action to remove or reduce the risk of those causes taking effect.

The chapter briefly reviews the possibility of organisational learning and looks at three broad levels of learning, highlighting that isomorphic learning may be the most valuable. The factors involved in active foresight are then discussed and the general concepts noted above are defined. A brief discussion of these concepts is followed by definitions of the types of organisations involved in disasters. The bulk of the chapter describes in more detail the eight concept groups and their relevance to particular organisation types.

5.1 Learning from disasters

A recurrent theme in this book is that while socio-technical disasters often appear superficially to exhibit a set of unique features, deeper analysis of the circumstances surrounding such incidents reveals the existence of similar patterns of behaviour across a wide range of such incidents. Such similarities have led Lagadec to conclude:

> ... the disaster must not be seen like the meteorite that falls out of the sky on an innocent world; the disaster, most often, is anticipated, and on multiple occasions.[1]

If disasters do recur for the same or similar reasons then it may also be postulated that the organisational learning which takes place following such events could also have similar features. Consequently, if we better understand those processes of reporting back and learning it may eventually be possible to design procedures and structures that will assist organisational learning and reduce further losses.

The research for this book was undertaken in an attempt to address this need for better understanding of organisational learning. The arguments propounded in this chapter and the model developed in the discussion of the research have been derived from the methodological approach discussed in Chapter 3.

5.2 Levels of learning

Learning from unwanted events can occur on at least three different levels of analysis. The first level is that of 'organisation specific' learning, where individual organisations involved in a particular incident each draw their own lessons from the event. The second level of learning is where, after analysis of the factors surrounding a specific organisational failure, more universally applicable lessons are drawn. This 'isomorphic' learning will be discussed in more detail later in this chapter. The third level is the most broad and general, and can be thought of as 'iconic' learning. This is the notion that simply being informed that a negative event, or some form of organisational failure, has taken place is a learning event in itself.

While the 'organisation specific' and 'iconic' levels of analysis and learning are important, it is arguably the 'isomorphic' lessons drawn from the analysis of a disaster that are the most valuable. At that level of analysis the underlying causes of complex large-scale socio-technical disasters can be seen more clearly. With the particular and individual circumstances stripped away, the core problems of an organisation's involvement with a disaster can be examined and compared to the activities of other organisations. Therefore, if strong isomorphic similarities are found, it may be possible to utilise the lessons learnt in those other organisations to help prevent recurrence of a similar incident.

5.3 Active learning

The evidence suggests that there are at least two types of learning that can take place. The first is 'passive learning', that is simply knowing about something; and the second is 'active learning', that is knowing about something and then taking remedial action to rectify the deficiencies that have been uncovered. Regardless of what lessons are said to be learned as a result of an inquiry into a disaster, unless those lessons are put into practice no 'active' learning has taken place. There is little point in knowing how to prevent a disaster if no active steps are taken to prevent it.

One tragic example of passive learning, as opposed to active learning, is the catastrophic failure of the space shuttle *Challenger* seconds after its launch. For some time prior to the accident it was known by engineers working on the project that there was a strong possibility that 'O' rings, which sealed the multiple sections of the solid fuel booster rocket, might fail under certain weather conditions and allow hot gases from the booster to escape. Unfortunately, active learning did not occur and as a consequence no positive action was taken to prevent the seals from failing. As a result, both the *Challenger* and its crew were lost when one of the 'O' rings burnt through during take off, releasing a jet of hot gas which ignited the shuttle's large external fuel tank.[2]

Another example of passive learning with disastrous consequences is that of the King's Cross underground station fire in London on 18 November 1987, when 31 were killed. More than two years before the disaster, in 1985, there had been a serious fire at Oxford Circus underground station. Following that incident the London Fire Brigade arranged a meeting at its headquarters with all those people who, it was believed, could make a significant contribution to increasing fire safety on the underground system. Of that meeting the [then] Chief Officer of the London Fire Brigade later noted that not only did he arrange for the senior management of London Underground Ltd (LUL) to attend but he also:

... included in that meeting not only principal operational and fire prevention uniformed officers but also the scientific advisor. [And] At that and subsequent meetings strenuous efforts were made by the Brigade to encourage LUL to call the Brigade to every suspected fire. *Unfortunately they resisted with equal vigour.*[3]

Similarly a 1988 scientific press article reports that London Regional Transport had:

...failed to learn the lessons of two fires on wooden escalators that broke out at Green Park station. Both incidents preceded last year's fatal fire at King's Cross station.[4]

The questions to be asked in all such cases are: did the information available reach the individuals or groups with the necessary authority to act upon it? and if so, did particular factors hinder or prevent action being taken? It was with such questions in mind that the research for this book was undertaken.

5.4 General and specific learning

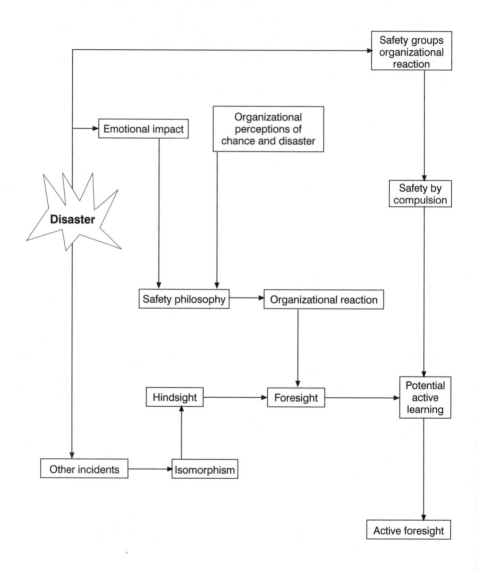

Figure 5.1 Steps to active foresight model

Analysis of interview data with senior personnel from organisations involved in disasters revealed that the lessons learned from such events appeared to take two distinct forms, general and specific. General lessons could be applied to any part of an organisation, whilst others were directed towards specific areas of organisational activity. In this chapter it is the former of these two sets of learning, the more general, that will be discussed. The more specific learning concepts will be discussed in Chapter 6.

To give some structure to this discussion it will help to look at Figure 5.1, which illustrates in a simple diagram the major relationships between the ideas discussed below and the way in which they can influence the development of 'active foresight' with organisations.

5.5 General concepts defined

The eight general concepts associated with organisational learning and noted earlier are defined as follows:

1. *Organisational isomorphism:* refers to occasions on which organisations or their sub-units, whether engaged in disparate enterprises or not, exhibit similar patterns of behaviour. The concept is derived from von Bertalanffy's notions of general systems theory discussed in Chapter 2.

2. *Emotional impact:* respondents' accounts of their own, and their colleagues', emotional distress at being involved in a disaster.

3. *Chance and disaster:* refers to the different organisational perspectives taken by senior personnel on the role of fate in the creation of a disaster in which their organisation was involved.

4. *Attempted foresight:* those beliefs held by organisational personnel about how foresight might be developed and about some of the problems which prevent it being used actively.

5. *Hindsight:* refers to some of the organisational views that were held on the use of, and problems associated with, hindsight.

6. *Organisational reaction:* this category relates to the way in which different organisations' attitudes are affected by a disaster.

7. *Safety by compulsion:* concerns the views of organisational personnel on the advantages and disadvantages of mandatory rules and regulations in relation to safety.

8. *Safety philosophy:* applies to the ideas which interviewees expressed when discussing matters relating to safety.

5.5.1 Discussion of general concepts

The following is a brief discussion of how it is envisaged that these general concepts might interlink. As can be observed from Figure 5.1, a disaster has an 'emotional impact' upon personnel which may affect the 'safety philosophy' of the organisation concerned. Another input which appears to play a part in determining the amount by which an organisation's 'safety philosophy' will change are the beliefs which organisational personnel have regarding the role chance played in the disaster. Following the disaster the then held 'safety philosophy' will influence the 'organisational reaction' to that particular event. The organisational reaction in turn influences the generation of 'foresight'.

The disaster itself can be viewed as an incident to be included among 'other incidents' which can provide opportunities for additional organisational learning through 'isomorphism'. The lessons generated through isomorphic learning then lead to 'hindsight'. The availability of 'hindsight' allows 'foresight' to be developed depending upon the extent and type of 'organisational reaction' to the incident. 'Foresight' in turn produces alternative opportunities for 'potential active learning' to take place. At this point in the diagram the foresight is passive.

A disaster additionally creates an opportunity for 'organisational reaction' from safety conscious pressure groups. This reaction often leads to an attempt to create new safety legislation, and enforce 'safety by compulsion'.

The generation of passive 'foresight' in combination with 'safety by compulsion' gives scope for 'potential active learning' to take place. This, when mitigated by organisational specific factors, determines the amount of 'active foresight' that will take place.

5.6 Types of organisations

Analysis of the disasters selected for this research suggested that at the highest level of generality the organisations concerned could be placed into six main categories. Every type may not be involved in each disaster, because different disasters affect and arouse concerns in diverse groups of people. For example, if a ship foundered at sea, the disaster would have a great deal of relevance for the general public if the ship were a passenger vessel. If it were a Royal Navy vessel, different organisations would be interested.

The six main types of organisation identified can be discussed under the following headings:

1. *Primary:* the organisation or organisations to which the disaster occurs. For example, the company that owned and operated the Summerland Leisure Centre on the Isle of Man which caught fire with the loss of over fifty lives. In some circumstances there may be more than one primary organisation involved. When a collision occurred on an automatic half-barrier railway crossing at Hixon, Staffordshire, between an express train and a road transporter carrying a 150 ton electrical transformer, there were two primary organisations involved. The first was British Rail, owner and operator of the train, and the second was Wynn Heavy Haulage, owner and operator of the road transporter.[5]

2. *Auxiliary:* organisations which have had some form of interactive contact with the primary organisation during the disaster's incubation period. Examples of such bodies in the case of a building fire would be the architect's practice involved in the original design or modification of the building, the contractors and sub-contractors who constructed the building, and the people and firms engaged to complete internal fabrication and fitting out.

3. *Alleviating:* organisations such as the fire, ambulance and police services. These attend disasters as a matter of course, since one of their fundamental roles is to give assistance at such events. In addition to these emergency services other organisations also fall into this category. One example is provided by the public health inspectors who traced people believed to have had contact with those who had been infected during the 1978 Birmingham Smallpox occurrence to ensure that they were quarantined in order to stop the spread of infection.

4. *Unionate:* organisations such as trade unions, or professional institutions. These will typically have no direct link with the disaster but are often required to attend an inquiry. For example, trade union officials may advise members who were involved in an incident, or a professional institution may be called in to give specialist advice on some particular aspect of concern. (The term 'unionate' is defined by Abercrombie et al.[6])

5. *Pressure group:* some of these have grown up informally as a result of a disaster, and then take on a more formal role. Examples include the Herald of Free Enterprise Families Association and Marchioness Action Group. Others, like the Royal Society for the Prevention of Accidents (RoSPA) exist formally prior to an incident taking place.

6. *Commissioning:* those organisations which commission inquiries. These are often, but not always, the government department responsible for overseeing the particular area of commerce or industry in which the disaster occurred.

These six types of organisation can be involved in a disaster at two physically and conceptually distinct levels. The first level is that part of the organisation which was physically involved with a disaster in some way. For example, the staff and management of the Summerland complex or the Coldharbour Hospital. The second level is more remote from the disaster: the board members of Trust House Forte, the company which owned and controlled the leisure complex, or the regional hospital committee responsible for the running of Coldharbour Hospital.

These particular examples refer to primary organisations, but a similar distinction can be drawn for each of the organisation types noted above. Each organisational type will be involved with a disaster on at least two distinct participative levels: (a) the on-site part of the organisation which is in close physical contact with the circumstances surrounding the disaster and its consequences; and (b) that part of the organisation which is both physically and organisationally remote from the incident but which will make the strategic decisions concerning what action, if any, should be taken in relation to the disaster.

This structure of the organisations associated with a major disaster is shown diagramatically in Figure 5.2. It should be noted that in some cases the boundaries between the organisation types is blurred, as for example with the Herald of Free Enterprise Families Association where many members of the pressure group were also actually involved in the disaster. In such cases both the perceived and real organisational distance from the disaster will be different.

Typically, for all organisation types, the distance that personnel are from a disaster appears to have a profound influence on the way they are affected by the incident, both in terms of its 'emotional impact' on them and of the degree to which an organisation's 'safety philosophy' may be modified. As well as influencing how people are affected, the relative distance which people perceive themselves to be from a disaster, when combined with the other factors, also seems to affect the amount of learning that they believe their organisation has to engage in.

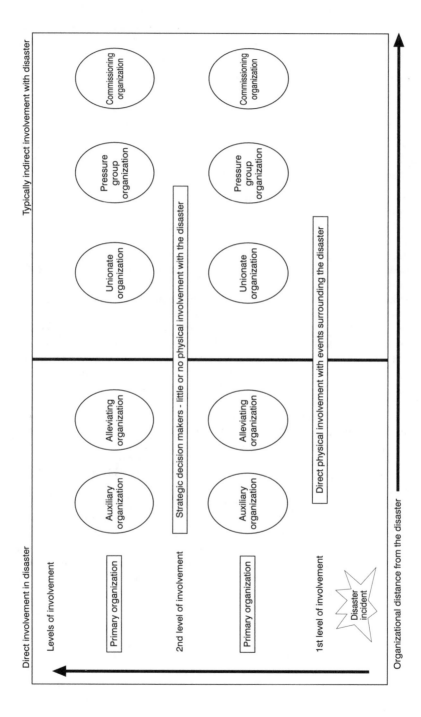

Figure 5.2 Organizational chart

5.7 Discussion of general concepts

5.7.1 Isomorphism

It is true to say that no specific accident ever occurs twice, for each individual disaster is unique. However, accidents do appear to have similar features at some levels of analysis. This empirical observation, when linked to von Bertalanffy's hypothesis on the nature of systems, raises questions about how far organisational learning can occur through the isomorphic features of an accident.[7]

There are at least four ways in which organisations can be viewed as displaying similar, isomorphic properties. The first, 'event isomorphism', is where two separate incidents take place and manifest themselves in two completely different ways but lead to the creation of identical hazardous situations. One example which illustrates this form of isomorphism involves an event which led to the rail accident at Clapham Junction on 12 December 1988 when 35 people died, and almost 500 were injured, 69 of them seriously. If a train driver passes a signal set at danger and so invades a section of track which already had a train travelling on it, this action is known as a 'signal passed at danger (SPAD)' incident. A second, isomorphic event, occurs when a train enters a section of track and the signal controlling access to that section fails to change to red and thus allows other trains to enter the same section of track at the same time. This is known as a 'wrong-sided signal failure (WSF)' since it leaves the signalling system in a dangerous condition.

Of these two events the Clapham Junction inquiry report states:

> Both these two breaches [SPADs and WSFs] can be caused either by equipment failure or by human failure on the part of the driver or signalman respectively if there is no automatic system to prevent human error. Looked at carefully from this viewpoint both ways are in fact but two halves of the same coin... Both have the potential to cause untold loss of life and damage, both should have received the same degree of attention from those involved in the management of safety in BR: they did not.[8]

The report continues:

> Nobody in management seemed to recognise that an unprotected wrong-sided failure would cause exactly the same risk of an accident. Nobody in management it seems grasped the nettle of the potential of the wrong-sided failure to cause an accident at least as disastrous as a SPAD.[9]

Thus, in failing to recognise the isomorphic qualities possessed by the SPAD and WSF events an opportunity was missed to prevent or minimise the risk of an unwanted incident.

The second kind of organisational similarity is 'cross-organisational isomorphism' which applies where the organisations belong to different sets of owners, are managed and staffed by different people, but operate in the same industry. While these organisations are dissimilar at one level of analysis in that they have different company names, locations and so on, at another level they can be thought of as identical as they each produce the same product or supply the same service in similar ways. For example, British Airways and American Airlines will be seen as quite different organisations by many people: they have different names, belong to different nations, represent different cultures and the values associated with them. They both, however, provide a very similar service using similar techniques and devices: that is, a service to ferry people around the world using aeroplanes. Thus both organisations will be open to a set of similar types of errors, failures and unforeseen organisational interactions that might lead to the creation of disasters.

One disaster which typifies the wasting of a chance to learn from this category of cross-organisational isomorphism is that of the crash of a Turkish Airlines McDonnell Douglas DC10 just outside Paris on 3 March 1974. A rear cargo door had been incorrectly fastened and subsequently due to vibration it had opened at approximately 12,000 feet with catastrophic results. The rapid decompression which followed the door opening caused the passenger cabin floor to collapse into the cargo bay and sever all the flying and engine controls which ran immediately beneath it. Without these control lines there was no way for the aircrew to retain control of the aircraft and as a consequence in that moment the passengers and aircraft were doomed.

However, this was not the first occasion on which a cargo door had fallen off a DC10. In June 1972, a similar set of circumstances to that just described took place aboard an American Airlines DC10 near Windsor, Ontario. On this occasion however, fortunately for all those on board, the aircraft was not carrying a full load of passengers. On this occasion when the cabin floor collapsed, the damage to the aircraft's control lines was not as severe as in the Turkish Airlines aircraft, and the flight crew managed to land the aircraft safely.

After the Canadian incident McDonnell Douglas had undertaken some remedial action, but it was, in the event, not sufficient to prevent a recurrence of exactly the same failure mode. The only difference on the second occasion was the far more horrific consequences of the failure.

It should also be noted that prior to the Turkish Airlines disaster different operators had filed to McDonnell Douglas over a hundred reports of problems regarding the closing of cargo doors on the DC10, many of them describing the same kind of faults. A fuller exposition of the Turkish Airlines disaster and many other major accidents that illustrate this type of organisational isomorphism can be found elsewhere.[10]

A more recent example of cross-organisational isomorphism is that highlighted by the Sandoz warehouse fire at Basle, Switzerland in November 1986 and a similar fire at an Allied Colloids warehouse in Bradford, England in July 1992. In both cases large quantities of fire-fighting water, contaminated with chemicals stored in the warehouses, found its way into the local water courses causing serious pollution.

Another way in which organisations can be seen to have isomorphic qualities, even though they appear to be of a disparate nature and in different industries, is through 'common mode isomorphism'. Here, although the organisations belong to different industries they use the same or similar kinds of tools, components, techniques or procedures in their operations. One instance of this form of isomorphism is the use of polyurethane foam in both aircraft and furniture industries for the creation of seats, sofas and other furnishings.

There have been many catastrophes where this material has been involved and the disasters have displayed this form of cross-industrial isomorphism. Cases which illustrate the concept are those of the fire which took place at Coldharbour Hospital on 5 July 1972,[11] the conflagration which took place on board a Lockheed Tristar in August 1980,[12] and the blaze aboard a British Rail sleeping car in the early hours of 6 July 1978.[13] In all three cases, in the considered view of the specialists brought in to ascertain the causes of the incidents, it was the flammability and toxicity of furnishings which were most to blame for the outcomes of the disasters. In each case polyurethane foam furnishings were singled out as the 'hidden' hazard. The same foam has been cited by the Fire Brigade and the Fire Research Establishment as being responsible for the speed of growth and the severity of the consequences of many domestic fires.

The fourth type of isomorphism is that of 'self-isomorphism' where the organisation involved is so large that it has many operational sub-units which generate or provide the same or essentially similar products or services. British Coal, British Rail, local government,

National Health Service hospitals, large industrial or commercial conglomerates like ICI or the House of Fraser provide examples. Many of the operational sub-units will be subject to similar internal and external contingencies and consequently to the same type of failure modes.

Although the managers who were interviewed in the course of this research did not use the terminology of 'isomorphism' they did appear to recognise implicitly that such a general principle might exist. One manager for example asked:

How do we make sure that what's learnt in one job is carried across into another?

And in relation to the concept of cross-organisational isomorphism as discussed above, one respondent noted that there were:

lessons to be learned which could avoid the recurrence of some of the crucial events in similar establishments.

Another respondent remarked:

Obviously we can learn from each other...as far as the railway is concerned we also try to learn from other industries.

Other interviewees' comments can be seen as relating to 'common mode isomorphism'. One manager recalled:

The disaster at Manchester Airport where the plane landed safely after failing to take off, and sixty people died in the tail of the plane and the evacuation time was two minutes. What killed them was not petrol – that started the fire. What killed them was the upholstery of the aeroplane.

While one other manager commented:

It [polyurethane foam] not only affected hospitals, it affected every public building in this country. The recommendations from this fire here – they were applied to nearly every public building whether it's a nursing home or whether it's a council office.

A recognition of 'self-isomorphism' can be seen in the responses by interviewees to the question: 'Did you use the information gained from the inquiry in any other part of your organisation?' Every single one of the organisations taking part in the research said that they had endeavoured to implement the lessons learned, where applicable, in other parts of their organisations. One respondent commented that:

we reviewed every single fire responsibility we had.

Another said that for:

any building, fire is something that's taken into account in every planning application that comes along.

And yet another respondent stated that:

what we did was double the night attendants at all residential establishments straight away.

The advantage of explicitly recognising where organisational isomorphism does exist is that this recognition allows the learning process to be considerably speeded up. If some form of socio-technical failure takes place in any organisation where isomorphism can be established, then the appropriate level of remedial action can be taken in those other organisations before they experience a similar failure. These other organisations do not have to 'reinvent the wheel' in order to gain the advantage of using it.

There is, however, a caveat to be observed when looking for isomorphic resemblances. Perceived similarities can be deceptive and this in itself can be dangerous. As one manager commented:

> you can have a similar looking event taking place when in fact it's not what is actually taking place.

Reason summarises this point when criticising the methodology underlying the concept of human reliability analysis pointing out that:

> error data do not 'travel' well from one situation to another.[14]

Fiering and Wilson also make the same point when they note that:

> Although it is conventional to assume that a risk calculated from historical data is reliable, this is only true if our model – that the future is analogous to the past – is correct.[15]

The report of the Coldharbour Hospital fire would appear to offer an example where one of the organisations involved did not realise that a situation existed which was not isomorphic with its own beliefs until after the disaster occurred. Regarding the organisation that issued furniture to hospitals, the report notes that:

> The policy of the department as one would expect provides for these types of articles being supplied throughout the country on a large contract and no special standards were imposed in this case.[16]

The organisation concerned supplied furniture to hospitals throughout the country and, typically, no strict standards of fire-proofing were thought to be required in any hospital since people in hospitals were presumed to be unlikely to start fires deliberately.

A further common assumption which the hospital authorities could have made with regard to a fire is that in most cases the alarm would be raised fairly quickly since any member of staff or any patient could bring to everyone's attention the fact that a fire had started. Such taken-for-granted factors did not apply, however, in the Coldharbour Hospital incident, for the furniture was supplied to a hospital for the mentally ill where in fact a patient did start a fire and did not raise the alarm. Thus there was a difference between the situation as perceived by the department issuing the item of furniture and the situation that existed in reality. This difference in perception ultimately led to the destruction of several hospital wards and the lives of a number of patients.

Organisational isomorphism should be looked for wherever it might exist for it is a useful abstraction which can greatly assist the learning process. But care should be taken to ensure that the situation which is to be learned from is genuinely isomorphic with that in which the lessons are to be implemented, so that the lessons learned can be safely transferred from one organisation to the other.

5.7.2 Emotional impact

The emotional impact on people involved in major incidents appears to be massive and enduring. This research highlighted one instance where the effect was so great that, even though the incident had occurred many years ago, a senior manager still did not wish to talk about it because:

> It still upsets me to think about it.

The organisation in which the person was employed was not in any way responsible for the incident.

A second example which again illustrates vividly the massive impact that disasters have upon organisational personnel, is the fire which occurred at the Fairfield old people's home in Nottinghamshire in December 1974. In this fire, over thirty residents lost their lives due to the inhalation of smoke while asleep. The Chief Architect of Nottinghamshire who was responsible for the design and maintenance of the home was so shocked by the event, that he spent the next ten years researching and designing a chair which would be comfortable for old people but which did not contain any polyurethane foam. The disadvantages of such foam in chairs is that it catches fire easily and gives off large volumes of toxic smoke. The chair which is now in production will eventually be introduced into all of the old people's homes run by the council and has been shown in tests to be safer than those previously purchased.[17]

Another demonstration of the effects that an accident can have on an organisation is the fact that some of those interviewed reported that their involvement with the accident had triggered a major shift in their preoccupations and activities. One architectural practice, for example, had shifted the emphasis of its work so that a sizeable proportion of it was now concerned with safety matters.

A further observation relates to the clarity of recall of incidents, even after an interval of ten years or more. There is no independent means of confirming the accuracy of respondents' recall. Studies of memory and recollection would lead us to expect systematic distortion in such retrospective accounts. However, informants' discussions of what had taken place had a very vivid and immediate quality. They had no difficulty in presenting a clear account of what had taken place, of the lessons that had been learned and of how the implementation process had been carried out. They had all clearly carried away and retained in a very accessible form their own personal lessons from the incident in which they had been involved.

The emotional impact on organisational personnel does also seem to vary with how intimately the person was involved in the disaster in terms of their 'relative distance' from it. These observations on how organisational personnel respond to being involved in a disaster closely correspond to those described by Taylor.[18] He suggests that the size and type of emotional trauma suffered by those associated with a disaster is a function of their relationship to it; that is, a function of whether the person is a survivor of a disaster, a relative or friend of someone directly involved, one of the rescue and salvage professionals who are brought into deal with the situation, or just someone who has read a report of a disaster in a newspaper.

Other evidence which appears to support the above view was presented at the 1990 annual conference of the British Psychological Society, in Swansea, by Dr. M. Mitchell, a psychology lecturer at Queen's College, Glasgow and Dr W. D. S. McLay, Chief Medical Officer of the Strathclyde police force.[19]

Their research consisted of a longitudinal study of the officers who had been involved in the police operation following the Lockerbie air disaster in which all 259 passengers and crew, and 11 people on the ground, were killed after a terrorist bomb exploded in a Boeing 747 in flight over Scotland.

Mitchell and McLay reported that one police officer who had patrolled the local golf course that night looking for wreckage and bodies stated that he was:

...quite certain that the memory of those body shaped craters six inches deep in the grass will remain with me for the rest of my life.

They also observed that the 190 police officers who had worked in the mortuary appeared to be even more traumatised than those who had patrolled the area searching for bodies. These findings seem to be consistent with the condition known as post-traumatic stress disorder.

5.7.3 Chance and disasters

As noted earlier the evidence suggests that it may be possible to prevent some disasters provided that action is taken sufficiently early. However, there is strong experimental evidence that people do not appear to enjoy learning from negative events even when the lessons gained through such learning situations may be advantageous to them.[20]

Furthermore, Langer in her paper on 'The psychology of chance' notes that there is a tendency for people to:

> ...attribute desirable outcomes to internal factors but to blame external factors for failures.[21]

This she suggested is because people wish to see all events as potentially controllable and therefore use attributional processes which will increase their sense of mastery if the event is a success and decrease their feelings of responsibility if the event is negative or a failure. By this means they can take credit for any success and avoid blame for failure.

Thus, if the outcome of a particular event is expected to be successful, and it is, then the:

> Initial expectation of success leads one to attribute the success of one's ability; if one fails then the failure may be attributed to bad luck, a result which makes the re-evaluation of one's mastery unnecessary.[22]

Another factor which may add to the problem reviewed above is that of the strength of a person's beliefs about the 'facts' surrounding a particular situation. For as Slovic and Fischhoff point out:

> ...a great deal of research indicates that people's beliefs change slowly, and are extraordinarily persistent in the face of contradictory evidence.[23]

Similarly, Janis in an earlier paper reported that in his study of group behaviour he found:

> Numerous indications pointing to the development of group norms that bolster morale at the expense of critical thinking... One of the most common norms appears to be that of remaining loyal to the group by sticking with the policies to which the group has already committed itself, even when those policies are obviously working out badly and have unintended consequences...[24]

Levitt and March also suggest that:

> ...stories, paradigms, and beliefs are conserved in the face of considerable potential disconfirmation; and what is learned (by an organisation) appears to be influenced less by history than by the frames applied to history... individual decision-makers often seem able to reinterpret their objectives or outcomes in such a way as to make themselves successful even when the shortfall seems quite large.[25]

Thus the amount of change that takes place in an organisation both physically and attitudinally after an accident may depend upon whether or not those managers in positions of power are able to change their views in the face of the new evidence or whether they believe it was just a case of bad luck that was at the root of the problem[26].

Some support for the above hypotheses was found during this research, as this comment from a respondent shows when asked if he thought bad luck had played a part in the creation of an incident:

> I think the feeling was, a year or two later, that it was very much a one off accident that had happened before, was very unlikely ever to happen again even if we hadn't taken any action.

This is in spite of the fact that prior to the disaster occurring two similar incidents had taken place which, even though nobody had been injured, were sufficiently newsworthy to have been reported in the press. Another respondent when commenting on the same question replied:

> Do you mean there but for the grace of God go I? Well yes I suppose that philosophy might have come into it.

One respondent expressed the opinion:

> I don't think any incident can be prevented with hindsight, there is no doubt about it.

However, most of the organisations were of the view that luck had played little or no part in the incident they were involved in. Responses by many managers in the organisations when asked if bad luck had played a part in the incident were not untypical of these respondents' comments:

> No, no, I don't think so.

One senior manager stated:

> There's no question of hard luck.

And another:

> I am inclined to think with hindsight that it could have been avoided.

However, many of the organisations did make the point that human fallibility was something which could not be predicted and therefore a random or chance element would always exist within the organisation. One respondent commented:

> People are human, failure will always occur.

And another:

> I think we must always be reminded of the human element and all the recommendations we care to make can never remove that weakness.

These views of the world echo the work done by Reason who argues that error prediction can only be made:

> ...from adequate models of human action.[27]

of which there appear to be none at the present time, thus suggesting that accidents from these types of events will not easily succumb to analysis and prescriptions for their avoidance.

5.7.4 Attempted foresight

Karl R. Popper in his work notes:

> ...even the greatest improbability always remains a probability, however small, and that consequently even the most improbable processes – that is those which we propose to neglect – will someday happen.[28]

Similarly Lewis notes that:

> When a great many rare events each have a small probability of occurrence, the chance that at least one of them will occur can be rather high. One often finds people saying that some event, which has extremely low probability, has occurred, so that there has been magic somewhere.[29]

Thus the foreseeableness of a disaster occurring arguably is dependent upon how the chance of that event taking place is evaluated in the first place. If an assessment of a risk is undertaken and no disaster occurs then, as discussed above in relation to chance, the people concerned with that evaluation will simply attribute the successful outcome to their ability to control the environment. However, if a catastrophe does ensue then one of the main bones of contention within an investigation is likely to be that of how forseeable the incident was. An example of this is noted in the following extract from the report of the Fairfield fire:

> There seems little doubt that had an automatic smoke detector system been in operation at Fairfield at the time of the fire Mrs Herbert and Mrs Johnson would have been alerted at a considerably earlier stage, and might well have got the Fire Brigade to the premises before the fire had reached serious proportions.[30]

Unfortunately for those concerned in such an event, psychological research indicates that, in hindsight (that is, knowing how things actually turned out):

> ...people consistently exaggerate what could have been anticipated in foresight.[31]

Thus the concept of foresight appears to be fraught with problems, not least of all for those who are supposed to have employed it.

It is clear, however, that many people in organisations do recognise the need to evaluate risks carefully and to take adequate precautions to prevent accidents from occurring, thus attempting to incorporate foresight in their operations. One of the central problems in attempting to use foresight is that in an open system we can never have perfect knowledge about all the variables with which we are likely to come into contact. Consequently the best anyone can be expected to do is to make 'reasonable' or 'rational' decisions based on the information available to them. Thus as Turner points out:

> ...if we are looking back upon a decision which has been taken, as most decisions, in the absence of complete information, it is important that we should not assess the actions of decision-makers too harshly in the light of the knowledge which hindsight gives us.[32]

This sentiment is echoed in the report on the public inquiry into the fire at the Fairfield old people's home. The Chairman stated that:

> '...of course we embark upon this [inquiry] with the benefit of hindsight and it must not be supposed that what is clear now would or should necessarily have occurred to anyone considering the situation before the fire.'[33]

However, since the evaluation of an organisation's foresight with the benefit of hindsight can lead to problems, decision-makers should, to protect themselves in case some untoward incident does occur, take the advice offered by Slovic and Fischhoff of making:

> ...a clear record of what they knew and the uncertainties surrounding their actions. Critics of these decision-makers might improve the acuity of their hindsight by attempting to state how events might have turned out otherwise or by seeking the opinions of persons not already contaminated by outcome knowledge.[34]

As already mentioned, within open systems it is impossible to have perfect knowledge and the best anyone can do is to make reasonable assumptions based upon the evidence available. However, there are times when the existence of a particular risk is known in advance and therefore the problem is not one of realising that a risk exists but one of deciding whether it is an acceptable risk to take. For example the report on the Fairfield old people's home fire states:

> some degree of risk from fire has to be accepted in homes for the elderly.[35]

This is because, the report goes on to suggest, excessive expenditure on fire precautions might have negative effects on homes for the elderly, existing homes having to close down if too stringent requirements were placed upon them. Similarly, the report on the Coldharbour Hospital fire points out that:

> ...we consider that the bedding supplied was satisfactory and that the fire risk involved by its use must be accepted and catered for in another way.[36]

This attitude suggests that in some cases even where knowledge is available it will not be used to stop the event taking place, but only reduce the chances of it occurring. Thus lack of foresight cannot be blamed every time a disaster takes place since clearly there are occasions when it has to be applied in a qualified manner.

Unfortunately, there are occasions when knowledge of a hazard is available to the management of an organisation, and that knowledge is not used at all, not even to reduce the probability of it creating a disaster. The following statement from one respondent shows this:

> One of the patrons had already warned the management about locked fire doors.

Nothing was done about this and as a consequence many people were injured or lost their lives in the fire simply because they could not get out. This type of behaviour is unfortunately still commonplace. In September 1991 a chain was found wrapped round the handles of an exit door at a London theatre, and the production company was subsequently fined for having locked exit doors.

Organisations do, however, in the main take steps to introduce foresight into their operations in a number of ways. Some of the methods related by respondents were: the use of project management techniques that employ a series of checks and balances to ensure that every aspect of the operation is supervised and managed efficiently; the use of long-term planning to ensure that the organisation changes in response to its environment, that appropriate internal monitoring operations are employed; and the institution of research programmes into areas where hazards are known to exist, since research will often provide, if not the immediate answer to a problem, at least a 'better' way of dealing with it. Finally, organisations may take some form of remedial action in response to warnings that a particular situation is hazardous. As noted above this latter form of attempted foresight is not always as effective in preventing disasters as it might be since it can be affected by factors other than those of safety.

5.7.5 Hindsight

One respondent noted above did not believe that a disaster could be prevented with hindsight, but many others did believe this as the following comments illustrate:

> I learnt my own lessons from that inquiry and what happened afterwards... You must be capable of learning from mistakes and recognise mistakes.

The usefulness of hindsight was demonstrated when respondents were asked: 'Has a repetition of the same kind of incident been prevented because of recommendations?' One manager replied:

> A second small fire was started in 1977 and it was dealt with very effectively within seconds.

A second stated:

I don't know. We haven't had another one [a fire] that's for sure.

And a third that:

There has been no repetition of an incident of that nature.

As with foresight, hindsight can be defeated by various means. For example the dangers associated with the use of polyurethane foam in soft furnishings have been known for some time. When one senior manager was asked why he felt that the foam was still being used in furniture he replied:

Because it's bloody great pressure groups supported by the whole of the British furniture industry and the oil companies and you are up against it.

Support for this notion that self-interested groups can affect the use of hindsight comes from a press report in May 1985 on the use of polyurethane foam in furniture:

By September 1980 it was clear that by simply calling attention to the foam hazard we were getting nowhere... Even the lighted match test was dropped after pressure from the furniture industry.[37]

As noted earlier, however, disasters and the inquiries which follow them create intense political and economic pressures which are often difficult for a government to resist, even if it has the will to do so. This is demonstrated by the following statement:

Two weeks after Woolworths [fire] Sally Oppenheim then Minister of State for Consumer Affairs, announced that the Government was producing new safety standards for furniture... This however was a fairly large red herring.[38]

Such comments thus reflect Behrens', observation that:

A disaster...only prepares the groundwork for change, it does not guarantee progress and in some cases does not lead to change at all.[39]

Another way in which hindsight can be thwarted, although not so deliberately as in the case of self-interested groups, is through the passage of time. As one respondent commented:

consciousness of the event dissipates over time

and as a consequence can leave the way open for a similar event to recur. This is particularly so for those who have no direct contact with the original disaster.

5.7.6 Organisational reactions

Whilst on some occasions little or no substantial progress is made towards a safer society in the aftermath of a disaster, on other occasions the opposite is true. For as Behrens points out:

A disaster creates a climate uniquely conducive to social reform and legislation.[40]

And thus one of the reactions to a disaster is for those organisations and groups of citizens who are concerned with safety to seize upon the incident and endeavour to use it as a lever to effect change.

One example of this form of reaction to a disaster can be seen in the way in which organisations like the Royal Society for the Prevention of Accidents and the British Safety Council in 1988, after a series of fatal household fires where it was determined that the composition of the furnishings had played a significant part in the tragedies, lobbied the government to introduce legislation to ensure significant changes to the permissible uses for polyurethane foam.

A second example of this form of response was seen following the fire at the Triangle Shirtwaist Company in New York on 25 March 1911. After the fire the outcry from reformers was so great that it led to the setting up of a factory investigating commission where the:

...legislature granted the Commission unparalleled discretion to conduct its investigation into state factory conditions.[41]

The Commission achieved many changes in state legislation including a law which completely reorganised the entire state's Labor Department and Labor Code, thus also effecting changes in areas not directly concerned with safety.

While most of the organisations whose members were interviewed for this research appeared to have reacted to the incident in which they were involved with a measured response, some organisations were thought by the experienced professional people interviewed to have overreacted. This can be witnessed in the following comments:

The authorities went beyond what they recommended in fact.

And :

They fail on the side of excessive safety.

However, while some organisations do overreact in their responses to disasters, it is not in their best interests to do so, for the implementation of excessively stringent or elaborate safety procedures can lead to organisational personnel endeavouring to circumvent them and hence nothing is gained from such a response. Similarly, the unconsidered spending of large amounts of money will not necessarily solve problems which led to a particular incident taking place. Those responsible for dealing with the aftermath of a disaster would do well to bear in mind the caveat given in the report of the Fairfield Home fire:

It would be unfortunate if in an exaggerated reaction to the tragedy all these measures were applied indiscriminately without reference to particular circumstances and regardless of cost.[42]

Thus an organisation's reaction to disaster should be suitable to the circumstances surrounding the situation if the most fitting solution is to be found and implemented.

5.7.7 Safety by compulsion

In the aftermath of a disaster there is often an outcry from different public bodies such as newspapers and safety organisations for the government of the day to enact legislation that will prevent such a disaster from recurring. As one manager pointed out:

At the end of the day, legislation that's the important thing.

The Fire Brigade have been campaigning for years to be allowed to set mandatory standards for hospitals and other public places which in Britain came under the umbrella of Crown Property. At the present time Crown Property enjoys immunity from prosecution and thus may set whatever standards they think fit regardless of good safety practice. As one fire officer pointed out:

we have no power to impose recommendations or anything else.

This situation exists, as one respondent pointed out, because of:

a total unwillingness by successive governments to give the Fire Service teeth with legislation, the necessary legislation.

The same person noted a short while later:

> The Home Office is very keen on self regulation. Now we think that's crazy for God's sake.

This sentiment is held by many safety organisations with the argument advanced that, as currently many companies and public bodies only pay lip service to the regulations, to allow them to control themselves would lead to a more unsafe situation. However, whilst mandatory regulations are useful as a tool to attempt to get people to behave in a responsible way, unless ways are found to create a culture in which 'regulations' are readily accepted it is doubtful they will ever be as effective as is intended. As comments from respondents below show, people will break the rules:

> Patients were not supposed to have cigarettes and matches in their possession but some of them did.

> A resident smoked in her room on a number of occasions before the fire when it was against the rules.

> The staff knew the rule [smoking in forbidden areas] was broken at other times.

While there is generally an effect from most legislation in setting a standard for law-abiding individuals or organisations, the presumption of law-abiding compliance of rules and regulations does appear to have an air of certainty about people's actions which is misleading. The idea that legislation will always be obeyed and hence particular types of incidents prevented from taking place is often not the case.

Rules can be made but if the people to whom they relate do not obey them then perhaps on occasions the situation is made even more dangerous than it was previously, since others within the organisation will perform their duties in the erroneous belief that particular procedures are being carried out when in fact they are not. Such rule breaking is often found to take place within the 'incubation period' of an accident as described by Turner.[43]

Legislation, rules and regulations all have their place in protecting society from the dangers which surround it. But it would be prudent for those in authority to consider that, at best, legislation and rules act as stern *aide-mémoires* to remind organisations of the operations they should perform and the standards they should attain, but they also may, at worst, act on occasion to create a sense of certainty where none may exist. Without physically checking every action of every member of staff there is no way of knowing if rules are being obeyed. The Chernobyl nuclear power station disaster in the USSR is a classic case of people disregarding rules which were laid down as mandatory.[44]

5.7.8 Safety philosophy

Each of the six main types of organisation involved in a disaster brings its own perspectives to bear when reviewing their own safety philosophy after being involved in an accident. For as Turner observes:

> the manner in which hazards are perceived will also vary to some extent according to the vested interests which those concerned have in different aspects of the dangerous situation; for example, nineteenth century factory owners would have seen the hazards of factory work and the precautions which might be regarded as 'reasonable' very differently from those working for them.[45]

One of the major factors that appears to relate to how organisations review their safety philosophy after being involved in a disaster seems to be the amount of organisational 'surprise' involved at the way in which events have turned out. Usually, the larger the surprise, the larger the shift in safety philosophy. For instance, the design, construction and operation of an old people's home may not normally be thought to involve a life-threatening situation. But it does, as the loss of life at the Fairfield old people's home demonstrates. Support for the notion of differential organisational surprise comes again from Turner who in a similar vein argues that:

> the inhabitants of Aberfan were apprehensive in a general way about the tips overhanging their village, but they would have kept their children away from school if they had been able to predict with considerable accuracy when and where the tips would slide. Their apprehension, however, allows us to speculate that they were less 'surprised', in some sense, than say the Chairman of the National Coal Board, who appeared to have had no such apprehensions about tips in general, or about Aberfan tips in particular.[46]

A comment typical of the organisations which had experienced a safety shock in this way when asked the question 'As a result of the incident was there any change in the organisation's safety philosophy?' was the respondent who replied:

> Yes there certainly was, yes certainly was.

Another respondent replied that:

> we were already conscious of safety matters but the incident sharpened everyone's interest and awareness.

Yet another stated:

> It's a lasting realisation that you can be very dangerous people.

However, on the other hand, if the everyday role of an organisation is to face life-threatening situations, as is the case for example with the Fire Brigade, then a change in safety philosophy is less likely to take place. Such an organisation deals routinely with other people's crises and disasters, so the mere fact of a disaster will not disturb them to the same extent, but provide them on occasions with an opportunity for fine tuning their procedures. The comment by one senior officer in the Fire Brigade illustrates this:

> We are continually honing and improving and tightening up all the time [procedure]...we are in an emergency service and it's our way of life, in other words it's second nature to us.

Thus, it is likely that very special circumstances would have to exist before such an organisation undertook a radical review of it own modus operandi: failure to cope with a new type of disaster would perhaps be one such circumstance, and another would be if a Fire Brigade were to suffer deaths from, say, food poisoning in its canteen.

The basic everyday activities of an organisation play an important function in creating the organisation's cultural expectations with regard to safety, and hence its perceived relative distance from a disaster. As a consequence, the most radical changes in safety philosophy as a result of a disaster will tend to come from the organisations whose cultural expectations have suffered the largest surprise.

However, the amount of change in an organisation's safety philosophy also appears to be dependent, particularly in the case of auxiliary organisations, on the amount of responsibility that the organisation feels towards the creation of the incident. For where an organisation does not believe that it was responsible in any way for the incident occurring, there will appear to them to be no need for a change in their safety philosophy. The following comment from a respondent typifies this attitude:

> Apart from the scrutinising of literature and the withdrawal of one piece of literature which whilst not intentionally misleading had an element of ambiguity. That was the only thing that was done and the only thing that we felt necessary should be done actually.

A senior manager from another organisation commented that:

> ...the inquiry didn't find us as consultants in any way negligent

and said that as a consequence there was no reason for them to alter their approach to how they managed their projects.

5.7.9 Change in safety philosophy

There do appear to be a number of factors which will affect the amount of change that can be expected to take place in an organisation's safety philosophy and, as a consequence, changes in its operational procedures or structure. These are:

(a) The size of the emotional impact on key organisational personnel; the larger the impact the stronger the compulsion to correct the mistake. Also the higher the level in the hierarchy that the impact occurs the greater is the likelihood that the changes will be implemented, since it is in the upper echelons that the power to bring about large changes resides.

(b) The type of organisation (that is, primary, auxiliary, alleviating...) and hence its relative distance from the incident.

(c) The degree of organisational 'surprise' it has been subjected to.

(d) The organisation's everyday role and cultural assumptions.

(e) The amount of responsibility it feels towards the production of the incident.

(f) Whether or not the management attributes the disaster to internal or external factors. That is, whether they believed that there were organisational techniques which could have prevented the disaster or whether just plain 'bad luck' was seen as the root cause.

However, as discussed earlier, it is doubtful if compulsion alone can change the safety philosophy of everyone. For whilst rules and regulations may change individuals' physical behaviour patterns towards some particular situation, it cannot necessarily affect their fundamental mental attitude towards matters of safety and that is where many of the problems lie.[47]

5.8 Conclusions

The analysis of the interview data discussed above suggests that there are a number of general factors which affect the way in which organisations learn from disasters. Some of those factors contain pathologies which can affect the amount of active learning that organisations might otherwise profit by. They are, however, seldom if ever explicitly on the agenda when people discuss how organisations can learn from mistakes.

The utility of the explicit recognition of these concepts is that much of the information contained within them can be readily incorporated into organisational training programmes, and, as a result, may lead to global changes in an organisation's safety culture.[48] These changes would lead to the creation of a climate within an organisation where the staff would be responsive to learning from hard experience and as a consequence improve both their and the organisation's chance of survival.

Notes

1 Lagadec, P. (1982) *Major technical risk: an assessment of industrial disasters*, Pergamon Press.

2 Presidential Commission, *US House of Representatives Report (1986), Investigation of the Challenger accident. (House Report 99-1016)*, Washington DC: US Government Printing Office.

3 Clarkson, D.G. (1989) *Safety management*, October.

4 *New Scientist* (1988) 25 February.

5 *Hixon report*, op. cit.

6 Abercrombie, N., Hill, S. and Turner B.S. (1984) *Dictionary of sociology*, Penguin.

7 Toft, B. (1992) Failure of hindsight, *Disaster Prevention and Management: An International Journal*, Vol.1, No. 3 November/December.

8 *Investigation into the Clapham Junction railway accident*, (1989) by Anthony Hidden QC, Cm 820, HMSO, London.

9 ibid.

10 Eddy, P., Potter, E. and Page, B. (1976) *Destination disaster*, Hart Davis, MacGibbon, London.

11 *Coldharbour report*, op. cit.

12 *Sunday Times*, (1980) 9 November.

13 *Report on the fire that occurred in a sleeping-car train on 6 July 1978 at Taunton in the Western Region British Railways*, HMSO, London.

14 Reason, J. (1985) Predicting human error in high-risk technology, *Lecture to BPS Annual Conference*, Swansea, March.

15 Fiering, M. and Wilson, R. (1983) Attempts to establish a risk by analogy, *Risk analysis*, Vol. 3, No. 3.

16 *Coldharbour report*, op. cit.

17 *Architects Journal* (1985) May.

18 Taylor, A.J.W. (1986) Models of disasters and human responses, paper invited for presentation to the Conference of the Society for Psychosomatic Research, 17/18 November, London.

19 *The Independent* (1990) 9 April.

20 Wason, P.C. (1960) On the failure to eliminate hypothesis in a conceptual task, *The Quarterly Journal of Experimental Psychology*, Vol. 12, part 3, pp 129-140.

21 Langer, E. (1980) The psychology of chance, in *Risk and Chance*, Dawie, J. and Lefrere, P. (eds.), Open University Press.

22 ibid.

23 Slovic, P. and Fischhoff, B. (1980) How safe is safe enough? in *Risk and Chance*, Dowie, J. and Lefrere, P. (eds.), Open University Press.

24 Janis, I.L. (1971) Groupthink, *Psychology Today*, November.

25 Levitt, B. and March, G. (1988) Organisational learning, *Annual Reviews in Sociology.* 14: 319-340.

26 Festinger, L. (1957) *A theory of cognitive dissonance*, Evanston, Row, Peterson, Illinois.

27 Reason, op. cit.

28 Popper, K.R. (1959) *The logic of scientific discovery*, Hutchinson.

29 Lewis, H.W., Budnitz, R.J., Kouts, H.J.C., von Hippel, F., Lowenstein, W., and Zachariasen, F. (1978), *Risk Assessment Review Group Report to the US Regulatory Commission*, The Nuclear Regulatory Commission, September, Washington DC.

30 *Report of the Committee of Inquiry into the fire at Fairfield Home, Edwalton, Nottinghamshire, on 15th December 1974* (1975) Cmnd 6149, HMSO, London.

31 Fischhoff, B. (1980) For those condemned to study the past: reflections on historical judgements in Shewder, A. and Fisk, D.W. (eds.), New directions for methodology of behavioural science: fallible judgement in, *Behavioural research*, Jossey-Bass, June. San Francisco.

32 Turner, op. cit.

33 *Fairfield report*, op. cit.

34 Slovic, P. and B. Fischhoff, op. cit.

35 *Fairfield report*, op. cit.

36 *Coldharbour report*, op. cit.

37 *Architects Journal* (1985) May.

38 ibid.

39 Behrens, E.B. (1983) The Triangle Shirtwaist Company Fire of 1911: A lesson in legislative manipulation, *Texas Law Review*, Vol. 62, p 319.

40 ibid.

41 ibid.

42 *Fairfield report*, op. cit.

43 Turner, B.A. (1978) op. cit.

44. Reason, J. (1987) The Chernobyl errors, *Bulletin of the British Psychological Society*, Vol. 40, July.

45. Turner, op. cit.

46. Turner, op. cit.

47. Toft, B. (1993) Behavioural aspects of risk management, AIRMIC Conference 1993, *New horizons for risk management & insurance*, Warwick University, April.

48. ibid.

Chapter 6

Specific organisational learning

Introduction

This chapter, as with Chapter 5, draws heavily on the responses from those interviewed for the research programme underlying this book. While the previous chapter looked at general factors affecting organisational learning, this chapter discusses seven concepts of more specific application. These concepts, derived from analysis of the research data, appear under the following headings: conditions for disaster; information and disasters; safety by regulation; personnel; organisational economics; organisational responses to public inquiry recommendations and lessons learned about fires. These specific concepts are factors which can affect the degree to which the potential for active learning is realised. The chapter begins with brief definitions of each of the seven concepts. It then proposes a diagrammatic systems model that highlights one way in which the concepts can affect an organisation's potential for active learning, and thus the amount of active foresight generated. It then discusses each concept in more depth.

6.1 Organisation-specific concepts defined

During this research respondents were asked about their experiences of organisational disasters, and about the learning responses of themselves as individuals and of their organisations. Analysis of the research data determined that, in addition to the general concepts discussed in the previous chapter, responses of a more specific application could be drawn into seven concept groups, each containing a number of contributory categories. These seven specific concept groups associated with organisational learning are defined as follows:

Conditions for disaster: respondent references to the conditions, events and issues associated with the period prior to a disaster.

Information and disasters: references to lessons drawn concerning the provision, use, storage, retrieval and transfer of information relating to accidents.

Safety by regulation: responses which referred to organisational factors associated with rules, regulations and codes of practice within socio-technical systems.

Personnel: references emphasising the supervision, training and recruitment of staff.

Lessons learned about fires: responses referring specifically to fire-related issues in socio-technical organisations.

Organisational economics: references to the variety of factors affecting financial and resource management decisions of organisations, both before and after a major accident.

Organisational responses to public inquiry recommendations: references to the organisational responses to both the recommendations of public inquiries and the implementation of those recommendations, including any factors that affected implementation.

6.2 Steps to active learning

Active learning is the process by which an organisation, after receiving information from a public inquiry, generates active foresight. To structure the following discussion it is useful to see how the various concepts outlined above may be involved in this process. Figure 6.1 is one way of illustrating how the different concepts relate to the learning process and to each other.

The active learning process described in the figure involves a feedback system operating in the same manner as the organisational control system outlined in the second chapter. Condition data on the state of an organisation are monitored by a 'perception of stability unit' (PSU) which is continually comparing the output of the socio-technical system (the organisation) against some explicitly stated criteria regarded by the organisation as being desirable. When a disparity between the system and the desired state is observed by the unit, an error signal, containing information about the nature and size of the discrepancy, is sent by the unit to inform a 'stabilising action unit' (SAU). The SAU then draws on available resources of the organisation and applies the control action required to drive the systems output back to the preferred level.

In this case the output is active foresight – the realisation of the potential for active learning. Figure 6.1 describes a system by which the amount of active learning undertaken by an organisation may be determined. Two external inputs to the 'potential active learning' system are the factors of 'foresight' and 'safety by compulsion' discussed in chapter 5. The internal inputs to the active learning control system are the seven factors related to specific areas of organisational life, as defined above.

Four inputs – 'information and disasters', 'safety by regulation', 'lessons learned about fires' and 'conditions for disaster' – can all be envisaged as inputs to the PSU. Each of these inputs contains lessons against which organisational condition data can be compared. For example, prospective organisational behaviour can be compared to see if it is about to violate any of the regulations created to prevent a recurrence of a disaster. Similarly, if one of the 'lessons learned about fires' is about to be compromised, error signals indicating the need for remedial action to prevent the violation can be communicated to the SAU. An example would include the identification of a fire hazard that might have arisen through the use of a particular material or procedure.

Two internal inputs – 'organisational economics' and 'personnel' – can each be thought of as inputs to the SAU. Each of these factors contains lessons which can be compared to any error signals emanating from the PSU. The output of the SAU would then be some form of action resulting in the organisation changing in some manner. For example, if the PSU detected that a difference existed from the preferred staff-to-client ratio, then subject to the factors applying to 'organisational economics', finance could be provided to enable more staff to be recruited. In the meantime lessons regarding 'personnel', such as a need to increase the level of supervision, could be implemented. 'Organisational response to public inquiry recommendations' can be thought of as the organisation's initial output response to the lessons which it has learned.

Now that the specific concept groups have been put into context, the rest of this chapter will involve a deeper discussion of each in turn.

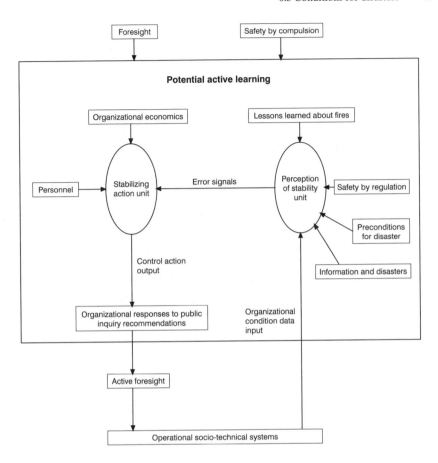

Figure 6.1 Steps to active learning model

6.3 Conditions for disasters

Although the interviews for this research were primarily conducted with regard to the post-disaster activities of an organisation, through this process the organisation's personnel also flagged up the conditions that existed prior to their particular disaster. This section will now discuss some of the events and conditions described by the respondents regarding their 'incubation periods'.[1]

6.3.1 Latent errors developed in the design stage

Turner suggests in his discussion of the incubation period of a disaster or accident that a propensity exists for unforeseen conditions to arise due to latent defects introduced into a system at the design stage. At that stage erroneous beliefs and assumptions may have already been made. For instance, clues and information regarding a system's condition, which the designer assumes will be available to those working in the system, may be masked by later actions. One senior manager highlighted this problem in a very physical manner when he commented:

It is very, very difficult to look at a wall that has been decorated and even hazard a guess as to its fire resistance.

In this case the decoration hides the physical clues to the fire resistance of the wall. A second example of the importance of design stage assumptions is that highlighted by the inquiry into the fire at the Fairfield old people's home. Referring to safety precautions, which might have saved lives had they been undertaken, the report noted that:

Once the home was built, modification presented problems which appeared almost insuperable to those who had to deal with them...The work could have been done at very little cost if it had been considered at the design stage.[2]

As well as being difficult to accomplish in some cases, desired modifications to existing designs may give rise to hazards unforeseen by the original designers. An insurance industry study group has noted this possibility associated with new designs or modifications:

If the machinery or process plant incorporates new designs or development features, this could present insurers with a greater potential for claims for breakdown than would be present in similar plant of tried and proven design.[3]

One apparent reason for hidden hazards arising after a system modification is that often at the modification design stage no strict specifications regarding the required change are given to the designers by those who commission the work. The specification is often fuzzy or is given verbally. The report into the Coldharbour accident notes that the consultants were:

...given a verbal brief which included the thinking and research which had already been done and in which the main thought was the therapy benefits of this kind of upgrading...That scheme and plan did not specify materials.[4]

In the Fairfield report, the committee noted that:

A number of standard drawings were available at the time and [the architects] made use of those they considered necessary.[5]

The report implies, therefore, that a very clear specification of the technical changes desired had not been given to those who were to design those changes. In such a case it is possible that, even before a project gets off the drawing board, an equivocal situation exists between the commissioning client and the consultant relating to the nature of the original brief.

6.3.2 Decision making in conditions of ignorance

While equivocal situations can arise because of inadequate briefing by clients, any fuzzy conceptions created in this manner can be amplified if the consultants themselves are not aware of crucial information. This does not necessarily come about through negligence. It may be that the emerging problem has not been anticipated by the consultants or anyone else in their profession.

An example is offered by an observation in the Fairfield report which notes:

It seems clear that the rapid movement of smoke and toxic gases through horizontal voids was not fully understood or appreciated in 1959.[6]

A related problem is reflected in comments on toxic fumes made by one respondent:

> The idea was not known at the time that smoke could creep along the corridors and kill people without disturbing their bed clothes.

Another respondent, who was professionally employed in the construction industry, pointed out that in one instance his organisation had no knowledge of what the client was going to do with a building after it was completed:

> ...thus making it impossible to envisage what problems of use might occur at a later date.

There are also times when information about a particular type of incident may be known but is not yet public knowledge. This prevents the widespread application of that knowledge to help prevent loss of life and property. As one respondent noted:

> ...they can't even...learn from the collective experience of insurance claims because nobody will divulge them.

However, as Turner points out:

> Knowledge about hazards, about precautions, about likely safety levels and about probabilities of different kinds of accidents is not evenly spread throughout society. Instead, such knowledge is distributed differentially between different individuals and groups...[7]

While it may appear that no learning is taking place, it might well be the case that the knowledge gained from accidents is restricted because it is to the advantage of certain groups to do so.

6.3.2.1 Interpretation

Misconceptions in interpretation of legislation and regulations may lead to unforeseen situations. An example would be the manner in which project designers interpret the text in documents such as Building Regulations. The committee investigating the Fairfield fire reported that:

> The bye-laws and regulations are complicated even when unearthed from the mass of other provisions of which they form part. We were left in no doubt by the architects who gave evidence that they found them confusing and difficult to interpret. The bye-law authority itself in our view misinterpreted the status of Fairfield as an exempt building.[8]

One senior manager echoed this view in a rather more colourful manner, saying:

> ...building regulations are written in legal language and we can't understand the bloody things.

Another manager alluded to this problem, when commenting on another organisation's interpretations of his company's advertising literature. He stated that:

> ...we had never interpreted it that way.

This particular problem of interpretation is also a topic for discussion below in section 6.5 on 'Safety by regulation'.

6.3.2.2 Organisational change

Another factor which can help to create or compound any ambiguity or fuzziness of thought is the evolution of an organisation over time. For example, one interviewee illustrated this very neatly when he stated that, as an architect:

> ...where you have to watch out is a change of use, if for example, a school becomes an Evening Institute.

Such organisational changes are important because the types of activities which subsequently take place may also change quite radically. In this case the change of use of a building to one for which it was not designed may place those who use it in hazardous situations.

The above is an example of a sudden organisational change. However, other changes can occur so slowly that they are either not noticed or are so difficult for personnel to notice that they are not formally reported. One manager commented that:

> ...what had happened was the situation had developed since the building was built.

Referring to the fact that over a period of years more old people had moved into the county, another interviewee noted that:

> Everybody was caught out by the demographic change.

The requirement to provide places for the elderly in residential homes, therefore, had unexpectedly increased. The same respondent also pointed out that:

> ...the average age had gone up from 65 to 85. You know it was virtually a geriatric hospital. We designed it as a friendly neighbourhood old persons home.

Clearly the needs of the residents, and hence those of the organisation, had changed over the years but no one had realised this. In a similar vein the report of the Fairfield fire inquiry points out that wooden furniture in the building was dried out by years of maintaining central heating at a high temperature. This had made the furniture and the building potentially more inflammable.

6.3.2.3 Complexity

In the immediate aftermath of a disaster the cause of the incident is often attributed to divine intervention, random chance events, technical malfunctions or some other simple causal mechanism. However, the public accident investigations that are undertaken frequently reveal a complex interwoven set of events which are socio-technical in nature. One example is the conclusion drawn by the inquiry into the fire at the Fairfield old people's home:

> The death of 18 people at Fairfield cannot be ascribed to any single cause. It will be clear from chapter 2 [of the report] that the coincidence of a number of factors was necessary to bring about the tragedy.[9]

A second example is an observation made in the architectural press:

> With most serious fires a number of events having occurred together make a tragedy inevitable and if one link in the chain is missing the fire may well be a minor one.[10]

A final example is taken from the Coldharbour report which states that:

one aspect cannot be divorced from others and the result must be a system incorporating checks and balances.[11]

It is not only the investigating committees and professional organisations which appreciate the complexity of disasters and the holistic nature of organisations. One interviewee stated that:

> ...there are so many things that can go wrong with buildings.

Another said that:

> ...one aspect cannot be divorced from others.

A further commented that:

> ...what one does can affect the other two or three.

One senior member of an architectural practice stated that:

> You must pay attention to the fire performance of the total structure and not merely the elements of the structure.

6.3.2.4 Demarcation of activities

Another problem, which can be considered part of organisational complexity, is the occasional failure to define the boundaries surrounding organisational activities. As the Coldharbour report notes:

> ...there is one respect in which responsibility is not entirely clearly defined: on the question of responsibility for deciding whether the doors in Winfrith Villa were to be locked or unlocked.[12]

Lagadec in his analysis of the Dudgeons Wharf fire in 1970 similarly notes that underlying causes of the disaster included:

> ...uncertainty, delays, communication problems...due to questions of demarcation.[13]

One of the respondents was of the opinion that the incident in which he had been involved could have been prevented if:

> ...the architect and the local building authority had been fully aware of their responsibilities.

One of the reasons for the emergence of the above problem was that operational procedures often evolved over time to meet particular local working practices. While such an evolution might be a satisfactory way to meet local requirements it can also lead to practices which are not formally specified, and thence to equivocal situations arising where it is not known who is responsible for what.

The evidence discussed above suggests that following a disaster the organisational personnel involved in its aftermath do not only learn about the activities and processes involved in implementing recommendations of an inquiry. They also gain a wider appreciation of some of the conditions and events which can, if not carefully monitored, be the precursors of catastrophe. As a consequence personnel may be 'better' prepared to deal with potentially hazardous situations.

6.4 Information and disasters

One of the key elements in keeping disasters at bay within any organisation is the existence and availability of necessary information. This data must be applied in the right place at the right time. It is through the flow of information that organisations attempt to co-ordinate their efforts to achieve optimum performance from a given set of resources. Hence, it is hardly surprising that where there are failures in communications, in some cases a disaster results. Here we indicate communication failures found to be of relevance in this research.

6.4.1 Mass media

One of the problems in trying to prevent a recurrence of a catastrophic event is that of communicating relevant information to all those individuals and organisations that require it. In all probability the first way in which most people receive information on any given disaster is through the mass communication media of newspapers, television and radio. Unfortunately, however, this form of dissemination of information is often more concerned with sales figures or audience ratings than with a dispassionate exposition of what has occurred. As one respondent commented:

> As soon as you ever get a railway crash in which somebody is killed it's banner headlines.

Another respondent related that:

> It was a bloody awful period of my life, bloody awful, television cameras all over the place and I appeared nightly on the news. OK the fall guy.

This aspect of media coverage has the unfortunate effect of making people very reluctant to advertise or to make known any failure in which they have been involved. One respondent made the point that:

> ...publicity of that nature does no one good at all.

Another interviewee, in relation to fire-fighting organisations publishing accounts of what they did when tackling particularly awkward or difficult fires, suggested that:

> You can say, human nature being what it is, they are not going to submit for publication an account of a fire that wasn't a success or near success.

The very information which is required to prevent a failure from recurring may, in many circumstances, never be brought into the domain of public knowledge because those who possess such information do not want to be publicly castigated.

However, there are occasions when the media has behaved in a positive way at the scene of a disaster. Scanlon has discussed the example of the fire, explosion and subsequent evacuation of 217,000 persons at Mississauga, Ontario, Canada, following the accidental derailment of a train comprising 106 cars, 39 of which carried toxic chemicals.[14] Scanlon notes that because the local police had established good relations with the representatives of the media, whenever a presentational or factual error occurred in media reports they were able to correct the mistake. Inaccurate or fanciful stories were not published, thus assisting the authorities to keep the population of the town aware of the current situation. Indeed so good was the media coverage that the local police:

> ...were flooded with letters after the incident praising them for their openness.[15]

If such goodwill and openness as that just described could be established on every occasion that a major accident is given media coverage, then society would be far better served than at the present time.

6.4.2 Public inquiry reports

Following a public inquiry a formal report is published containing a description of the events leading up to a disaster, an analysis of those events, and recommendations aimed at preventing the recurrence of a similar incident. However, due to the ad hoc way in which public inquiries are commissioned there appears to be no standard practice as to whether the organisations involved in the inquiry see a draft copy of the report before it is published. A Home Office spokesperson who had been involved with the Hillsborough football disaster commented that, as far as he was aware:

...there was no formal provision for draft copies of a report to be sent to the organisations involved.

Nor would it appear to be routine that the organisations involved in a public inquiry are sent a copy of the final report, as the following comments of respondents indicate:

I think we bought ourselves a copy, I think we went round to HMSO.

I have a feeling we might have been sent one.

We saw a draft I think.

Similarly, there appear to be no formally established procedures by the administrators of courts of public inquiry (the Treasury Solicitors) for ensuring that professional bodies which might have an interest in a particular disaster are informed that the report in question has been published. Given the importance of the final report to the organisations involved in the incident and the wider general public this is a less than satisfactory state of affairs. The assumption that it is enough to place the report in the public domain may not be sufficient in contemporary society.

Having received the report of a public inquiry, from whatever source, an organisation must disseminate, act upon and store the information contained within the report so that it will be available for future reference. When respondents were asked how, or even if, they had recorded the knowledge from an inquiry, 60% of them indicated that it was simply by keeping a copy of the report itself.

The remaining 40% said that extra information had been obtained from other sources such as professional journals and from their own internal investigations. The report and the additional information were in documentary form and had been placed in their archives. In no single case had any of the organisations used computer technology to record the information, either at the time or subsequently, even though on many occasions the organisation was using computers in its business operations.

6.4.3 Dissemination of inquiry information

The methods used to disseminate the information contained in reports throughout the organisation take a variety of forms. In the first instance an organisation would often arrange for circulars, or the report itself, to be distributed. Where appropriate, hazard warnings were also supplemented with signs or notices giving details of what action was to be taken in given circumstances. One respondent commented that:

> The doors at Fairfield were all posted with a notice that they should be kept closed during the stated hours.

A second stated that:

> ...there are now guidelines on every ward which are placed very prominently where staff can see them in relation to what action to take if a fire occurs.

6.4.3.1 Departmental conferences

In addition to the use of circulars, some organisations, particularly the smaller ones, made special arrangements for departmental conferences to be held to discuss the issues which had been brought to light. In large organisations like fire brigades or nationalised industries, where arrangements already exist for frequent interdepartmental and safety committee meetings, the information would be placed on the agenda and discussed along with normal business.

6.4.3.2 Revision of information

Where appropriate, instruction manuals and other existing documents used in the organisation were revised in the light of an inquiry's findings. Also where applicable, new rules, regulations and codes of practice were drawn up to promulgate the information and implement the recommendations made. In several cases the organisations involved had their own technical libraries which were supplied with information from the public inquiry and other sources. However, it should be remembered that records in any form can get lost or thrown away. One respondent noted that:

> ...we have no records relating to that period because they had been destroyed some years before.

6.4.3.3 Effects of organisation size

In organisations such as the utilities or nationalised industries, the process of disseminating information may be difficult because of sheer size. However, they do possess a bureaucracy which is in direct communication with every operating unit. Thus, within a short space of time the findings of an accident report can be spread throughout the organisation. The latter is typical of what normally occurs. The individual large organisation with several physical sub-units to which the recommendations relate can also disseminate information in this manner. Therefore, if an industry consists of a small number of large organisations, it is likely that at the very least the information will be distributed to, and implemented in, more than one location even if other organisations within that industry do not pick up the information.

The problem of information dissemination is much greater when a small organisation is involved in a serious incident within a large fragmented industry consisting of numerous organisations. As a general rule, organisations involved in an incident will use the information made available by an inquiry. However, other organisations in the same industry do not appear to make contact with an organisation that suffers a disaster, either to enquire about the conditions and events that led to the incident or to ask about the efficacy of the inquiry recommendations. This could be because all the organisations to which the information would be relevant have read the inquiry report and acted upon its recommendations, or because they have 'picked up' the information from another source such as professional literature or the press. Equally, it could be because the other organisations in the industry have not noted either the original incident or the report, and therefore do not know of the existence of the information.

Whatever the reason, none of the organisations in this research which had been involved in a disaster inquired of other similar organisations to see if they made themselves cognizant of the events surrounding the incident. Conversely, nor had any organisation requested information from those who had been involved in a disaster about the nature of the event or the remedial work which had been carried out.

Similarly, it would appear that often where a small single organisation had been involved in a disaster, the lessons generated by the inquiry were only put to use in the location in which the incident originally occurred. This is despite the fact that many other comparable organisations were subject to the same risk.

6.4.3.4 Exchanges of information

Some interorganisational exchange of information does take place but typically this is carried out by large organisations such as British Rail which send copies of their incident reports to the rail organisations in other countries. Similarly, information is also gathered and distributed by bodies funded through the taxpayer such as the Fire Research Station, Central Fire Brigades Advisory Council and the Health and Safety Executive.

Information from public inquiries is also disseminated through professional institutions. One respondent noted that one of the activities in which he had been engaged after a disaster had been:

> ...talking to the Institute of Fire Engineers and we talked to various branches [of the Institute] as well.

Besides arranging for lectures to be given to their members, professional institutions also disseminate information through the journals and proceedings which they publish. Such journals reach quite a wide audience, as can be observed in this comment from a manager who said that:

> The Fire Protection Association of course did a great deal of work on this [the Coldharbour fire] and their publications go out to architects, civil engineers and so forth.

Journals are also used by organisations to supplement formal reports of inquiries. A senior manager in one architectural practice commented:

> ...an architect is expected to have read the Fairfield Inquiry, read everything else on the subject.

Professional institutions can play a greater role through the publication of journals specifically containing reports on accidents from those organisations and individuals involved. A prominent example is the *Loss Prevention Bulletin* produced by the Institute of Chemical Engineers. However, while professional journals are a very useful source of information it should be noted that for individuals or organisations to receive the benefits contained in such publications they do have to have access to them. This costs money, and in many smaller organisations it may be a cost which they cannot bear, or are not prepared to.

6.4.3.5 Organisational memory and people

Perhaps one of the best information storage and retrieval devices that an organisation possesses is its staff. One manager pointed out that recommendations are:

> ...inculcated into the minds of the people who are involved.

Another respondent, asked if there was some form of system to ensure the lessons are not forgotten, replied that:

> We have a machine here: the County Architect is in charge of the whole thing.

When asked what would happen upon his retirement the same respondent said:

> Somebody else better than me will take over...the organisation is not just one man deep it has 350 staff who also know the story as well as me and took part in the inquiry and it's an ongoing thing.

6.4.3.6 Loss of organisational memory

Practically all of the informants claimed to remember with clarity the events surrounding a disaster, the lessons learned, and the implementation of inquiry recommendations. This suggests that if a second similar incident does occur to the same organisation, it is not because the personnel involved in the first incident have forgotten but because the organisation's memory has failed. That is, the lessons have not been disseminated or translated so as to ensure that they will remain within the safety culture of the organisation, industry or society.

Tentative support for this notion comes from a paper by Kletz which describes how a leaking gas isolation valve led to an explosion which killed one man. One of the recommendations from the subsequent inquiry was:

> Never place absolute reliance on a gas-holder valve, or any other gas valve for that matter. A slip-plate is easy to insert and absolutely reliable.[16]

Kletz then describes how, four decades later, a fitter was dismantling a large pump to repair to it. As the workman removed a cover, hot oil spilled out and caught fire because the suction valve had been left open. Three men were killed as a result of the error. Following the inquiry into this fire instructions were issued, one of which was:

> The equipment must be isolated by slip-plates (blinds) or physical disconnection unless the job be done so quick that fitting slip-plates (or physical disconnection) would take as long and be as hazardous as the main job.[17]

Kletz notes that:

> In the period of 40 years that elapsed between these two incidents, the practice of slip-plating had lapsed – no one knows when or why.[18]

He then asks the question:

> Will we forget again?

Clearly, this form of organisational forgetfulness[19] is also related to the problems associated with incremental change which have been discussed earlier. People die, leave work or change employers for many different reasons and in doing so the organisation's 'soft' human memory changes. Many respondents readily appreciated this form of organisational forgetfulness, either implicitly or explicitly, with comments like:

Recommendations tend not to be so easily forgotten if it's equipment. But people change and priorities change and therefore it is possible that things will be missed out because of such social changes;

and,

You can say that people who have been through here and have passed on to other places have taken a bit with them.

6.4.3.7 Advice ignored

Other information failures occur because staff do not act on the information they have been provided with. One respondent pointed out that immediately prior to one fire the hospital involved had:

partially ignored the advice of the Fire Brigade.

On another occasion a hospital that was redesigning some of its wards did not act on the advice of a hospital design note. Had it done so the fire which took place might never have turned into the inferno that it did.

6.4.3.8 Poor internal communications

Often problems arise because information does not reach the people who require it. In some cases, organisations do not have any formal machinery for passing information on and there is little or no exchange of information between members of staff. Where organisations do have such a formal arrangement, staff often do not attend the meetings which have been arranged. As one interviewee stated:

Communication between the Night Nursing Officer and the Senior Nursing Officer wasn't very good because he [the Night Nursing Officer] never attended the meetings.

On one occasion, this led to the emergency services initially being given inaccurate information on their arrival at the scene of the fire. According to a senior fire officer:

There were people who came down to assist at the fire who worked in that place and didn't even know that the place was occupied.

In fact the building concerned contained over thirty mentally ill patients.

It is possible for organisations to learn actively from disasters and yet, within a short period of time, because of human beings and their characteristics, to 'forget' those lessons, that is unless organisations can fashion ways to ensure the lessons learned from old incidents are transferred to new personnel. One of the ways in which organisations try to do this transfer is through specification of the behaviours organisational personnel are to engage in – through rules and codes of practice – and this is now discussed.

6.5 Safety by regulation

From interview responses, it is clear that organisations involved in disasters acted quickly to implement the recommendations of subsequent inquiries. However, as we shall see below, there were problems of implementation when large amounts of capital expenditure were involved. It takes time and money to test, develop, construct and incorporate new technology into any organisation. On many occasions, while organisational learning had taken place there was an unavoidable time lag between a particular lesson being learned and it being implemented. In the interim, one of the ways in which a shortfall in safety was made up was through production of new rules, regulations or codes of practice and then through drawing the attention of staff to those rules. In some cases the production of new rules and procedures was considered by the inquiry to be the main form of action required.

Sometimes, however, an accident had occurred not because rules, regulations and codes of practice did not exist but because the previously existing set of rules had not covered the conditions which had created the disaster. As one manager commented:

...the Theatre Regulations in the Isle of Man were not the same as Building Regulations and perhaps it should have passed through our minds that this difference was a potential hazard.

Another remarked that prior to the accident:

...there were of course Codes of Practice for hospitals long before this.

Quite often, the change in organisational behaviour required, was not brought about by the creation of completely new sets of regulations but by changes to the existing rules or codes of practice.

6.5.1 Interpretation of regulations

The interpretation of written or spoken information by different human beings is not unproblematical. Since the environment in which any organisation exists changes over time, the situation can be made even worse, particularly if rules, regulations and codes of practice do not keep pace with those changes.

Organisations then have problems in interpreting the 'old' regulations in terms of the new situations. As one respondent noted:

There is always one of those occasions that whichever set of criteria you react to you cannot win.

Another interviewee, when speaking of one architectural publication, said that:

...the Office Design Guide section on fire hasn't been revised since 1975.

The report of the Fairfield fire noted that:

There remains in relation to departmental notes and bulletins the danger that older publications may tend to fall out of line with newer ones written in the light of more recent research.[20]

Similar problems are created when a building to be constructed or altered does not appear to come under existing legislation. In this case there are often problems in interpreting the rules in the new circumstances, and consequently on some occasions a project will be exempted from complying with specific regulations because it is thought that they are not applicable. An example of this can be found once again in the report of the Fairfield inquiry where the committee note that:

The building regulations impose only minimum standards which in some cases – and Fairfield is one of them – do not cover every set of circumstances.[21]

A second example came from a respondent who remarked that:

> Unfortunately it is not possible in fact to design, say, even a small cinema, leave alone a centre of this kind and still comply with the regulations and therefore, there are certain waivers incorporated in every building.

A safety industry journal illustrated this problem when it reported that:

> The Health and Safety Commission has wrongly withheld information on industrial hazards from the public for the last eleven years because – as it now admits – it misinterpreted the law.[22]

A further possible problem with written procedures involves the definitions which may be attributed to the words used. As one respondent noted:

> You see, a fire officer will interpret the regulations one way and another one from another area interpret it another way.

A professional journal for architects also made the same point when it reported that:

> In the building regulations a fire stop is inaccurately defined as a barrier to the spread of fire or smoke instead of and smoke.[23]

Yet another problem surrounding the production of appropriate rules, regulations and codes of practice is that in general they are designed to meet a typical situation rather than an unusual or exceptional condition. The report on the Coldharbour fire, when discussing why a particular document did not cover certain aspects of the project, notes:

> This [document] is a note which concerns all hospitals of every kind and the writer did not envisage the special requirement of Coldharbour.[24]

Similarly, in the case of the Fairfield fire the committee noted that:

> It seems clear that the note was referring to general emergencies in relation to residents' care rather than fire.[25]

6.5.2 Review of existing rules

One of the strategies used to prevent the recurrence of a disaster is a review of existing practices at regular intervals. However, even when such reviews are undertaken such a strategy does not always work effectively. Prior to the Fairfield fire, a comprehensive survey and review of all the building regulations relating to building stocks was held by the County Council. While the strategy of reviewing regulations is undoubtedly 'good practice', managers employing such tactics should not be lulled into a false sense of security by its use, for this procedure alone cannot prevent major accidents from occurring.

6.5.3 Limitation of rules

Similarly neither rules, regulations nor codes of practice in themselves make any organisation disaster proof. The Fairfield inquiry report states:

> Building bye-laws and regulations should be recognised as laying down only minimum standards.[26]

Indeed, the fallacy of believing that rules can prevent all manner of misfortunes is clearly spelled out in the Fairfield report when it states that:

...some architects tend to rely on regulations and available guides, rather than on their own understanding of principles, to design against fire risk.[27]

One senior manager acknowledged this problem explicitly. When asked about rule following he replied that:

...they had attached too much credence to the regulations... Since the inquiry architects in his office are expected not only to have read all the necessary literature but also, on the subject, he uses his own brains.

Another manager commented that:

...as a professional you are not supposed to do as you are told. You are supposed to think.

This supports the notion that there is no substitute for people using their skill, experience and intelligence to solve organisational problems.

6.5.3.1 Rule breaking

To complicate matters still further, organisational personnel often find, as one respondent pointed out, that:

Routine actions become boring and a nuisance after a while and as a consequence habits are formed which break or circumvent the rules of the organisation.

After one incident a senior manager pointed this out, observing that:

there were a number of them [sleeping car attendants] found asleep or not doing the job properly.

Similarly, another respondent reported that a fire officer whom he knew would go round an office block removing wedges from fire doors, but:

...five minutes later they would be back in place.

thus defeating the purpose of safety doors and creating a situation that should not exist.

Often in the wake of a disaster the cry goes out that some form of legislation is required to prevent a recurrence of the incident. However, it is clear from the above discussion that such faith in rules and regulations can sometimes be misplaced. While statutory legal devices are useful they are not the panacea which many people believe them to be.

6.6 Personnel

One of the areas of organisational life which frequently comes in for criticism in the aftermath of a disaster is the standard of training of personnel. The report of the Coldharbour hospital fire comments:

In the hospital as a whole the proportion of unqualified nurses is too high.[28]

Later the same report states that:

...probably additional training of existing staff is needed to carry out these duties adequately.[29]

A respondent highlighted the same problem when he noted that:

> There was a need for better qualified staff,

in his organisation prior to the disaster.

The research highlighted two main methods by which organisations responded to the above problems. The first was to take on extra staff with the appropriate qualifications. A senior manager in one architectural practice illustrated this response when he stated that after an inquiry:

> We took on a Fire Prevention Officer....to advise on every drawing.

The second response to the exposure of poor training was for organisations to change or alter their training methods. One respondent commented:

> The accident resulted in a completely new piece of machinery and a different training regime for the men who operate it.

In-house training was also used to ensure that the lessons learned from an inquiry were incorporated into the organisation's safety culture. Every one of the organisations involved in the research stated that, where possible, the lessons from an inquiry had been adopted in the appropriate organisational training schemes. While the production of rules and regulations does not necessarily ensure that the organisational personnel will change their behaviour patterns, they appear to have a more powerfully beneficial effect where they are incorporated into formal training schemes.

6.6.1 Passive and active training

In relation to matters concerning safety, the research suggests that staff training can be of two distinct yet interelated kinds. The first form of training is where the staff are told to read, or have explained to them, sets of rules, regulations or procedures. This can be considered to be 'passive' training. The second type of training is where employees are given a period of passive training and then exercised in that activity to test out their understanding and their ability to behave in the prescribed way. This kind of training can best be described as 'active' training. It is clear from the research that active training is the method preferred by those organisations which have been involved in major accidents. As one manager put it:

> No amount of reading regulations is sufficient in itself. It's what comes after that, that's the important bit.

All the organisations interviewed appeared to be aware of the importance of appropriate training and the need to keep abreast of the changing needs of their organisations. This awareness can be seen illustrated in comments such as:

> There is a joint training committee...which discusses training methods and continually updates them in the light of new experiences from whatever source;

and,

> We are continually all the time improving and tightening up our training.

6.6.2 Supervision

Supervision within organisations is undertaken for many reasons: to direct and guide workers in their tasks; to stop workers from engaging in unproductive behaviour such as talking to each other about non-work related activities; or even, as in prisons, to ensure that prisoners do not escape. However, regardless of the context, essential elements of supervision are those of inspection, direction and control. By the use of such techniques, organisations endeavour to co-ordinate the different activities into a coherent set of complementary actions that will fulfil organisational goals.

The supervisory function can be thought of as the action by which someone or some mechanism oversees organisational behaviours. Where those activities do not correspond to organisational criteria, some form of control action will be instigated to stop the unwanted actions. Supervision and its associated supervisory action can therefore be thought of as being elements of an organisation's negative feedback loop, in that when an organisational activity is perceived which is away from some specified norm, appropriate control action is instigated to counter the perturbation and restore the organisation's state to equilibrium. Thus where adequate supervision either does not exist or fails to monitor dysfunctional behaviour, and where self-reinforcing positive feedback loops of adverse activities develop, they may eventually severely harm or even destroy the organisation.

6.6.2.1 Behaviour of supervised

When questioned about the role that supervision or the lack of it had played in the production of an incident in which their organisation had been involved, many respondents frequently made reference to the expected behaviour of those who were being supervised as the main criterion for establishing the kind and amount of supervision provided. For example, one interviewee made the comment that:

It all depends upon the level of understanding of the residents.

Another said that:

It all depends upon age, physical attributes, sex, age etc., of the residents.

A third respondent pointed out that:

It helps a great deal in an understaffed old people home if you can give all the patients a lot of sleeping pills.

Where organisations are dealing with the general public then discretion has to be exercised in the supervision of their patrons. Similarly, supervisors cannot run roughshod over the employees under them without bringing about undesirable behaviour in those workers.[30]

6.6.2.2 Supervisor-to-supervised ratio

The expected behaviour of those to be supervised will have an effect on the supervisor-to-supervised ratio, particularly where the visual sighting of the supervised is difficult because of environmental factors such as partitions and solid objects between the supervisor and the supervised. A respondent noted:

That type of construction, that type of layout was all right from an environmental point of view, but from the point of view of vision of the nurse, he couldn't see what was going on.

If the assumptions about the behaviour of those being supervised are erroneous, or if these assumptions or the environment change over time then the organisation may eventually have an inappropriate staff-to-client ratio. The evidence from catastrophic incidents suggests that when they take place the ratio of supervisors to supervised tends to be too small. This appears to be the case in the Fairfield and Coldharbour incidents. This comment from the manager interviewed illustrates the point:

> One person on night duty was not sufficient to look after the ordinary needs, let alone crisis needs of 56 varying degrees of bed ridden old people.

6.6.2.3 Supervisory failure

One clear example of supervisory failure is to be found in the report of the committee appointed to investigate the circumstances surrounding the use of contaminated infusion fluids at Plymouth General Hospital in 1972 which resulted in the loss of five patients' lives.[31] The investigation found that contamination of the infusion fluid had occurred as a result of the bottles failing to reach the required sterilising temperature at the manufacturer's plant. There had been a failure by the autoclave operator to follow the operating instructions and this had resulted in the contaminated batch being produced. The report then goes on to note that:

> The Committee considers that no real responsibility lies with him; it was for his supervisors to check that the operating instructions were carried out.[32]

Another example of supervisory failure can be found in the Taunton railway sleeping-car fire where, upon investigation, it was discovered that some of the attendants were sleeping at their posts. Similarly regarding the Coldharbour fire where the investigating committee noted that the night nursing assistant was absent from his post for a much longer period than they considered acceptable and that the supervising charge nurse should not have allowed this.

Unfortunately, there are many other incidents which could be used to illustrate the dangers to be found when adequate supervision is not provided by an organisation, but the case studies mentioned above adequately reflect the terrible consequences of such failures. When respondents were asked about how they dealt with this type of problem the typical method was through the use of extra training.

6.6.2.4 Safety programmes

The final way in which an organisation's management may try to increase supervision over possible hazard areas is through the creation and continual updating of safety programmes. However, it should be remembered that these programmes also rely on organisational personnel to carry them out and so care must be taken when making assumptions of how effective they are at minimising and spotting possible risks to the organisation.

6.7 Organisational economics

All organisations, no matter how large and powerful, have a finite amount of resources. Whether an organisation is commercial, public or charitable there is always a need to find economies and to decide priorities for the available resources. The report of the Fairfield inquiry noted this dilemma of resource priority and allocation when discussing the need for fire precautions:

> The needs of the elderly are in competition with many other kinds of need for scarce public, private and charitable funds. Excessive expenditure on fire precautions, particularly in existing Homes could result in fewer new Homes being built, and some existing Homes might have to close if too stringent requirements were imposed on them.[33]

The role which finance plays was also recognised by respondents, as this comment illustrates:

> Where the problem did lay, and of course we did recognise this although we didn't say it, was the cost involved.

Another respondent pointed out that his organisation had to consider:

> ...not only the safety but also the financial implications of the work to be carried out.

Another stated that:

> We used to attend a Fire Advisory Committee every month until it was closed down by Mrs Thatcher's [the then Prime Minister] cuts.

Due to the limited resources at their disposal organisations always have to engage in trade-offs, however much they may dislike doing so. A respondent demonstrated this point:

> There are so many demands on our money. By and large we want money going into patient care. I mean, I know fire precautions is patient care but it's hidden.

Similarly, another interviewee said:

> It will often be a question of comparing the cost of wages for extra staff with the cost of installing and maintaining a smoke detector system.

A third said:

> You can only afford so much manpower into a given topic of the day... What we are trying to achieve is a balance.

One of the problems in trying to achieve what is believed to be the optimum financial solution to a problem is that it can later turn out to have been a false economy. For example, prior to the fire at the Summerland leisure centre, in order to prevent patrons already in the complex letting their friends in free, some of the fire exits were locked. This prevented people from being able to evacuate the building as quickly as possible and as a result added to the general confusion and possibly the toll of injuries.

A second example of false economy was the small economy made during the construction of an old people's home. A senior manager related:

> As one of the minor economies made at the start of the contract to save a little bit of money, we cut out draught excluders from bedroom doors, which were put there to exclude draughts not make people lives safer. It was a nice little touch in an old people's home it was valued between 250 and 350 pounds (sterling). We cut these things out as indeed we probably should have, looking for minor economies and frills. It is quite

possible had we not saved that money that 18 people might have lived to a riper old age.

The absence of the draught excluders had permitted the toxic fumes from the fire to spread more rapidly and easily into the bedrooms and to the occupants.

Organisations try to economise on their expenditure. However, in the aftermath of a major accident the post-disaster costs of rectifying matters appear to be of little consequence to many organisations. Respondents reported that consideration of costs had not led to any recommendations being either modified or dropped in their organisation. In a similar way, all respondents were sure that most if not all of the post-disaster recommendations which applied to them had been implemented.

This attitude, particularly by primary disaster organisations, is illustrated by the following comments from respondents when asked about the costs of the recommendations. One interviewee replied:

> There was no problem, the money was voted additionally. Money for the protection of residential establishments.

Another manager said:

> I think it would be true to say, in matters of this nature, finances is not a point of issue.

A third replied that:

> ...money is no object where life and limb comes in.

It would appear that the organisational criteria for assessing appropriate costs levels change radically following a disaster. Arguably, had a post-disaster attitude to costs prevailed prior to a disaster, the organisations involved might never have found themselves in such a situation.

One respondent, however, did report that:

> We had a joint sort of Management Committee to get the rebuilding going which was chaired by Trust House Forte's (the primary organisation) and the Chairman once referred to this being a contract out of control.

A second interviewee said that:

> The County Council, like everybody else did, started putting unnecessary things in and spending a lot of money.

While the majority of those involved with financing the remedial work were not overly concerned with the costs incurred, a small number of individuals thought that the balance had moved too far in the opposite direction. However, when asked if the substantial costs involved in the rectification work had meant a cost in terms of job losses in their organisation, not one respondent reported that this was the case.

Questioned about the cost incurred by their insurance companies and whether their premiums had been increased as a result of the incident the respondents replied that they had not. In fact one interviewee reported that in his organisation:

> ...the premium rates had in fact come down!

However, this respondent did note that in recent years the insurance market had changed considerably and that the overall cost of insurance coverage was going up.

While organisations generally do not appear to worry too much about the costs they might incur in the aftermath of a disaster, there are always economic limits to what an organisation can do. The following comment by a respondent illustrated the point very clearly:

There might be a recommendation which whilst it would reduce the risk of an accident to zero would cost hundreds of millions of pounds and which they couldn't do.

6.8 Organisational responses to public inquiry recommendations

The possibility of any recommendation for change, be it commercial or safety, being implemented within any organisation depends on a number of factors. One is the level at which the recommendations are discussed within the organisation. In each of the case studies in this research the recommendations from the public inquiry have been discussed at the highest level, and support was given to them at this level. This resulted in the organisations being able to respond quickly, since the required resources could be immediately allocated by those with authority to do so. We might expect that if the recommendations had been discussed at a lower level of management first, it would have taken some time before any action could be initiated. In some organisations, the recommendations were discussed at all levels.

Earlier, it was pointed out that while organisations were often prepared to finance inquiry recommendations without a quibble and would have implemented them instantaneously, it was, however, in many cases impossible for them to do so because of organisational, technological and environmental constraints. The implementation of recommendations can take time, so that an organisation is still at risk of having a repeat incident in the interim period. The time scale can be quite large, as one respondent pointed:

> We have 3000 buildings and lots of people in our care. It meant through the coming months a careful review of everything.

That process was just to identify where the recommendations applied, not to start the remedial work.

However, some recommendations were implemented quickly. These typically were those which referred to desired changes in organisational procedures, training and recruiting of additional staff, and notification of particular hazards to both employees and public. One respondent noted that:

> Some measures were taken on the same day as the fire, due to what had been found during the course of the internal inquiry.

While another manager noted that fire-fighting lectures had been implemented:

> ...very quickly after the fire.

When asked if there had been any spin-offs from the implementation of the recommendations, two of the respondents commented that there had. The first said that:

> We got much more interested in how people use buildings and whether in fact smoke stops or fire doors were wedged open with newspaper and that sort of thing.

The second respondent said:

> ...work on signalbox-train radio telephones...This arose out of Taunton.

Usually, the person appointed to oversee the implementation of the recommendations was self-selected by virtue of their post within the organisation. This in turn ensured that those who were to manage the changes in the organisation had sufficient organisational power to order the changes required. A senior manager also made the point that:

Normally the person implementing the recommendations would carry on with their normal work. But you could get the situation where somebody was taken out of their job and do a particular task, but it's most unusual.

Typically, the person appointed to implement the recommendations would periodically have to make reports to his superiors on how the work was progressing, thus helping to prevent the organisation from forgetting about recommendations that could only be implemented over a long time.

Very few of the organisations had a change in their structure following a disaster. The few changes of a structural nature that were made consisted of creating new specialist posts to deal with safety-related matters, fire precautions in particular. One respondent stated:

Fire precautions became very much more important in the Health Service and saw the appointment of Fire Officers for Health Authority Areas which I think before 1972 were probably non-existent.

Another said that:

A person was appointed to be a supremo whilst others were given the task of reporting to him.

The vast majority of respondents, when asked if their organisational structure had changed in response to the recommendations replied, 'No'. One manager replied that:

In principle we didn't change.

Another explained that they had not changed:

...the structure of the practice or the relationship between partners or job architects and service engineers or anything like that.

While not many organisations had to change their organisational structure, all of them did have to alter their operating procedures or create new ones as a result of inquiry recommendations. Procedural change – either alteration of existing procedures or creation of new ones – can be placed into one of four categories. The first relates to the changing of supervisory and inspection practices, as this comment from a respondent graphically illustrates:

Nobody informed me that the age and range of infirmity of the residents was changing significantly. They bloody well do now I have made sure of that.

Another respondent noted that:

There was a tightening up of procedures with regard to the examination of the literature before it was published. Previously it had been produced by the sales organisation and while they relied on information supplied by the technical organisation the draft document was never scrutinised critically by the technical organisation before it was printed.

The second category of procedural change refers to the production of new or changed training procedures. Examples include the fire-fighting lectures for nurses noted earlier, and the encouragement of staff to voluntarily undertake extra qualifications.

The third category is change in the use of administrative procedures. These include cases where check lists were designed, recording procedures tightened up, information circulars produced, codes of practice, rules and regulations drawn up or altered, and periodic reviews of operational procedures detailed.

The fourth category involves changes to physical procedures such as current operational or maintenance practices. These are specific to each of the organisations concerned. For example, the operational changes that sleeping-car attendants had to make to their working practices are physically different to changes made in a drawing office.

In order to ensure that a particular type of incident does not recur it is important that the recommendations become part of the changing fabric of organisational life. One of the ways in which organisations endeavour to do this is by incorporating the lessons learned into the organisation's culture and thereby institutionalise the recommendations. This is illustrated by the following comment from a manager who was asked if there was any method for ensuring that the lessons learned would not be forgotten:

> They've been sort of inculcated into the culture if they were not in the culture already before.

Another respondent remarked that the legislation and the procedural changes made to the organisation are one way in which the recommendations:

> ...are automatically included into organisational culture.

A third manager said that:

> The procedures are the system that ensures the recommendations are not forgotten over time.

Once instituted, the changes made to an organisation were monitored by procedures that had been designed specifically to check that they were being adhered to. For example, one manager reported that:

> Each year every nurse should have a fire fighting lecture which is recorded and that list is passed on to the fire officer.

Not a single respondent reported that the recommendations which had applied to them had not proved effective. Moreover, when asked if, with hindsight, they wished to drop any of the recommendations, none of the respondents said they wished to do so. One interviewee stated that a few years earlier there had been a small fire which had been put out very quickly, and he attributed this to the fire precautions and procedures which had been instigated as a result of inquiry recommendations.

While it is clear that all organisations can learn from catastrophic events, it was pointed out by several respondents that they felt there was a difference between how small and large organisations learn from particular incidents, and that this could be a source of organisational learning difficulties. One respondent said:

> Learning from an unpleasant experience must vary very much between individuals and large organisations.

This statement is echoed by another manager of a small organisation who said in response to a question about whether or not he thought the recommendations which applied to his organisation had stopped a recurrence of the type of incident which had occurred:

> The complicated part of the question is that in our kind of general practice we are not dealing with a string of similar buildings, with similar purposes. If our practices consisted of designing old people's homes obviously the answer would be yes but since neither before or since Fairfield have we designed an old people's home, we haven't much scope for answering your question.

Whilst modifications to an organisation can be far more readily fed in where repetitive actions are concerned there is still the general organisational learning difficulty noted by

Levitt and March:

> The sample size is particularly acute in learning from low probability, high consequence events.[34]

These problems of low sample size and organisations not doing repeatable work are precisely the cases where the identification that a particular project has isomorphic properties with another one which has failed in the past can be of value. Even though the organisation might not have been involved with that type of project before, the collective experience of all those who have had analogous failures would be available to help them not repeat the mistakes of the past.

6.9 Lessons learned about fires

The focus of the investigation in this book has been primarily with fires and explosions, and the lessons taken by respondents from these specific types of disastrous events will now be discussed. The first lesson which respondents reported was that a fire can spread at an incredibly fast rate once it has been ignited. The applicability of this observation by respondents is supported by Beckingham who in a paper on disastrous fires in places of entertainment notes that:

> The recent fire in a club in America where 86 people lost their lives is another clear indication of the rapid speed of fire.[35]

The speed of a fire, however, depends to a large extent, upon the combustibility of the material which is being consumed and as a consequence it is wise to use materials which are either non-flammable or fire resistant wherever possible.

A second lesson is that once a fire has started, raising the alarm and getting the building evacuated can be problematic. In three of the five disasters covered by the interview programme the fires took place at night when people were asleep. Also, in not one of those fires, according to the evidence, did the staff of the organisation raise the alarm. In each case staff were notified about the fire by a resident or patron using the system after they had been roused from their sleep by smoke. In such situations it would seem that there is a need for remote sensing smoke detectors.[36]

Having raised the alarm, evacuating the area near to the fire can be a further problem especially where the structure or area concerned has a complex interior layout. As one respondent commented in relation to the layout of the building in which he worked:

> I was concerned about it (evacuating the building) because it appeared to be a maze. It was difficult to get out and even I (a member of the staff) had to ask the way out of that particular part.

Problems arise particularly where there are 'strangers' (as Turner[37] terms them) to the organisation. Strangers in this case can be thought of as those persons who are just visiting that part of the organisation and therefore are not familiar with its geography or its procedures. Difficulty can also be experienced in evacuating people if they are disabled in some way. Evacuation procedures need to be tested 'live' to ensure that they work and that staff know what to do. Similarly, the way out needs to be clearly marked, preferably on the floor, since smoke rises and obscures signs which are high up.

Smoke is a major killer in fires, with people often dying in their sleep as a result of smoke inhalation. The report of the Coldharbour fire notes:

The patients who died were already dead or unconscious through asphyxia and carbon monoxide poisoning before they were burnt.[38]

This constitutes a further compelling argument to have smoke detectors fitted and to ensure that fire doors are kept shut to prevent the spread of smoke.

6.10 Conclusions

Organisations are complex, ever-changing entities which exist in a world where perfect knowledge is impossible. It is therefore not possible to predict every outcome from all decisions made during the course of an organisation's life. However, research can identify a number of elements – conditions and events – which may damage or destroy an organisation if allowed to occur. It often seems to be the case that before organisational resources are released to combat such threats, a disaster has to occur. Organisational learning and change are welcomed in the aftermath of a disaster but appear to be resisted prior to it.

The evidence from respondents in this research suggests that where the lessons from disasters are implemented in organisations, the risk of having a similar disaster recur is significantly reduced. This suggests that if those lessons were also implemented in organisations which have the same or similar features to that which suffered a disaster, then those organisations could also be spared the horror of being involved in a similar catastrophe.

Notes

1 Turner, op. cit.

2 *Fairfield report*, op. cit.

3 *Report by Advanced Study Group 208A* (1986) Construction and erection insurance, Insurance Institute of London.

4 *Coldharbour report*, op. cit.

5 *Fairfield Report*, op. cit.

6 ibid.

7 Turner, op. cit.

8 *Fairfield report*, op. cit.

9 ibid.

10 *Architects Journal*, (1975) 1 October.

11 *Coldharbour report*, op. cit.

12 ibid.

13 Lagadec, op. cit.

14 Scanlon, J.T. (1989) Toxic chemicals and emergency management: The evacuation of Mississauga, Ontario, Canada', in Rosenthal, U., Charles, M.T. and Hart.P.T. *Coping with crisis: The management of disasters, riots and terrorism*, Charles Thomas, Springfield, Illinois.

15 ibid.

16 Kletz, T. (1980) Organisations have no memory, *Loss Prevention Technical Manuel*, Vol.13.

17 ibid.

18 ibid.

19. Kletz, T. (1993) *Lessons from disaster: how organisations have no memory and accidents recur*, I Chem E.

20. *Fairfield report*, op. cit.

21. ibid.

22. *Safety Management* (1965) October.

23. *Architects Journal* (1975) October.

24. *Coldharbour report*, op. cit.

25. *Fairfield report*, op. cit.

26. ibid.

27. ibid.

28. *Coldharbour report*, op. cit.

29. ibid.

30. Benyon, H. (1973) *Working for Ford*, Allen Lane.

31. *Report of the Committee appointed to inquire into the circumstances including the production, which led to the use of contaminated infusion fluids in the Devonport section of Plymouth General Hospital*, (1972) Cmnd 5035, HMSO, London.

32. ibid.

33. *Fairfield report* (1975), op. cit.

34. Levitt B. and March, op. cit.

35. Beckingham, H. (1985) The safer hat for managers, *Safety Management*, May.

36. *Smoke alarms in the home* (1990) National Housing Town Planning Council, May.

37. Turner, op. cit.

38. *Coldharbour report*, op. cit.

Chapter 7

Case studies

Introduction

This chapter contains case studies giving brief descriptions of the events leading up to five disasters. It is important to remember that these studies were written with the benefit of post-disaster research, and with knowledge of inquiry investigations and report contents. The studies are not an attempt to give a complete description of the circumstances surrounding the events – fuller commentaries are contained in the various inquiry reports. The case studies are included here to put the current discussion into context by looking briefly at some events that can occur during the incubation period of a disaster. The descriptions highlight the fact that while technical factors are involved in disaster causation, social factors – including communications failures, interactions between people, economics, and politics – also play a major role. The five disasters are as follows:

- Coldharbour Hospital – a fire at a hospital containing mentally ill patients
- Dudgeons Wharf – a fire, and subsequent explosion, during demolition of a storage tank
- Fairfield Home – a fire in a residential home for elderly people
- Summerland – a fire at a large-scale leisure centre
- Taunton railway – a fire in a train's sleeping-car

7.1 Coldharbour Hospital

In the early hours of the morning of 5 July 1972 a patient at Coldharbour Hospital, an institution for the mentally ill, set fire to flammable material in a ward. The fire brigade was alerted by staff and responded quickly and efficiently. However, once started the fire caught hold rapidly. The considerable amount of combustible material in the ward, as it burned, gave off large volumes of carbon monoxide and smoke. Thirty patients died – the post-mortem report stated that all of the deaths were the result of asphyxia and carbon monoxide poisoning.

7.1.1 Summary of the incubation period

The events leading up to the catastrophic fire at Coldharbour Hospital commenced with the desire of both the regional hospital board and the hospital's management committee to improve the quality of life of the patients in their care. It was decided that one method of achieving this aim would be to provide living accommodation that was less institutional in nature. The rationale underlying this proposed change was twofold. First of all, there was simply a wish to improve the patients' physical living conditions. The hospital wards had originally been designed to accommodate injured naval personnel during the Second World War and were of rather a spartan and 'barrack room' appearance. The second motive for change was the hope that a more 'homely' atmosphere would have therapeutic benefits and assist patient recovery.

Initially three wards were upgraded to produce a more domestic setting. The results were so encouraging that the hospital's management committee, with the full support of the staff, decided to go ahead with substantial improvements. Following consultation between the various interested parties a sketch plan and proposal document, outlining the planned improvements, were produced and sent to the local fire brigade for advice. After considering the proposal the brigade's fire prevention officer wrote advising the regional hospital board's architect that only '0' class fire-resistant standard materials should be used for the various partitions, walls, ceilings and space dividers.

In the meantime, the regional hospital board's assistant architect had familiarised himself with the contents of the Department of Health and Social Security (DHSS) document 'Guidance for Hospitals Design Note No.2', and had reached his own conclusions as to how the advice of the fire prevention officer should be interpreted. Once the board's architect had acquainted himself with the fire brigade's advice, and the regulations relevant to the task in hand, he believed that the materials proposed for the improvements would be satisfactory for their intended purpose. At this time hospitals were not required to obey the advice of the fire brigade because of Crown immunity, and the board's architect had wide powers of discretion regarding which materials could be sanctioned for use. The architect was able to, and in the event did, authorise the use of materials of a lower standard than that advised by the fire brigade. It should be noted, however, that while all those concerned with the project were aware of the fire risk problem their overriding concern was with the provision of a more therapeutic environment.

In August 1970 a consultant architect was called in for discussions on the proposed improvements. At his verbal briefing the emphasis was on the project's therapeutic nature. Although at this meeting the consultant architect was given both the sketch plan and the up-to-date documents relating to the project, he was not informed of the fire brigade's advice regarding the class of materials for the wall, partitions, etc. While later giving evidence at the public inquiry into the fire the consultant also stated that the idea that someone would always be attending to the patients had influenced his thinking in the design of the improvements.

Having received his brief the consultant architect inspected the wards that had already been partially improved. A meeting in October 1970, at which the consultant architect and a representative of the fire brigade were attending, did not discuss the class of materials to be used in the project. In due course the consultant drew up plans for the improvements, which were submitted to the regional hospital board's architect and approved.

In the plans submitted for approval not all the partitions specified were of class '0' standard material. There were several reasons for this. First, the consultant had not been informed of the fire brigade's advice. Second, the consultant's brief did not specify the use of class '0' standard material. Third, the consultant did not envisage that the divisions in the new wards would be structural, but rather regarded them as being part of the furniture and fittings.

At a meeting on the site in July 1971, held to discuss the material specifications, an assistant of the consultant architect expressed some disquiet over the advisability of using plasterboard (a class '0' material) in the construction of the ward improvements since it might be vulnerable to kicking and other dysfunctional behaviour from the patients. It was agreed between those present at the meeting that hardboard (not a class '0' material) should be substituted for the plasterboard. At a subsequent meeting between the consultant architect and his assistant, the consultant agreed to the proposed substitution of materials believing that it would not reduce the fire standard to less than class '1'. In tests carried out after the disaster at the Joint Fire Research Station, Boreham Wood, Herts, it was found that this change in materials, coupled with the type of covering fabrics used, had reduced the fire class in one case to class '2' and in another to class '4'. The report of the public inquiry also noted that:

These alterations which inevitably meant a reduction of the materials so far as surface spread of flame was concerned, were never discussed with the Fire Authority and their advice was never asked.[1]

The work on the buildings was finished in the spring of 1972 and accepted by the regional hospital board as satisfactory. It was then the task of the DHSS Supply Department to provide the furnishings for the new wards. Since nobody had informed the supply department that the patients could be fire risks, it supplied standard furnishings rather than materials with a high resistance to fire. Thus the wards were constructed and filled with materials having a fire resistance standard considerably less than that originally advised by the fire brigade.

The patients who were to occupy the new ward were unfortunately a mix of both low and high risk categories. The low risk category consisted of those patients who were not considered to be a risk to the public. The high risk category patients had threatened to, and in some cases already had, set fires. The reason for mixing the two categories of patients together was quite simply because of a shortage of humane accommodation for the high risk patients. It was also thought that, while the arrangement was not ideal, the risks associated with the dilemma of supervision versus therapy were acceptable.

During the early hours of the morning of the 5 July 1972 both the charge nurse and the nurse supervising the patients went for a hot drink together, leaving the patients in the Winfrith Ward unsupervised. It should be noted that the nurse in charge had not been given any detailed instructions, written or otherwise, regarding what procedures should be adopted by night duty staff. Nor was there any evidence to suggest that the night nurse should not leave the ward to make a hot drink.

During the time the nurses were absent from the ward a fire was started by a patient. Because of the inflammability of the materials used in the ward improvements the fire rapidly gained hold. Eventually a patient raised the alarm by shouting for the nurses who immediately operated the nearest fire alarm, thereby alerting both the fire brigade and the rest of the hospital.

When staff from other wards arrived at the scene a number of problems immediately beset them. First, very few people present had any idea of the numbers and locations of patients in the improved wards. This was largely because of poor communication between the night staff and the rest of the hospital personnel, and between the various night staff themselves. The main reason for this poor communication was the difficulty that night staff had in attending day-time meetings which were held while they were asleep. A second problem facing the hospital's staff during the crisis was that most of the external doors on the ward were locked to prevent the high risk patients absconding. Another difficulty, related by a respondent although not mentioned in the inquiry report, was that:

Each of the locks on the Winfrith ward were different.

As a result, several individual keys were required to open the doors. Furthermore, while staff had received fire precaution and fire-fighting instruction there was no formal evacuation plan for them to follow. Hence, attempts to evacuate the patients were not organised and as a consequence not as effective as they might have been.

It is clear from the events surrounding the disaster at Coldharbour Hospital that even good intentions can lead to unforeseen adverse consequences. However, if there had been an adequately tested crisis management plan to deal with such a situation it is quite possible that the consequences of the fire might not have been so severe.

7.2 Dudgeons Wharf

On 17 July 1969 a fire was inadvertently started during demolition of a disused tank farm at Dudgeons Wharf in East London. An oxypropane cutting tool, being used to dismantle an oil product storage vessel, ignited a pocket of inflammable vapour within a tank. The fire brigade was called to assist with the incident, but while it was attending to the fire the tank exploded. Six people lost their lives.

7.2.1 Summary of the incubation period

By 1967 the company owning the Dudgeons Wharf tank farm had no further use for it. Unable to find anyone interested in purchasing the property as a going concern the owner elected to sell the land for development. To facilitate this it was decided to demolish the tank farm and clear the site.

Among those putting in bids to carry out the site demolition and clearance were two large and experienced companies. However, the contract was given to what was effectively a one-person contracting company. The contractor was a trained and skilled welder and coppersmith but had no previous experience of large-scale demolition.

By the time the demolition contract was to be awarded the condition of the site had deteriorated to a semi-derelict state and was full of refuse. The two larger contractors would have required the tank farm owner to clean the site prior to demolition, or to pay for cleaning. These prospective contractors would also have required assurances from the site owner that the tank vessels were thoroughly clean inside, or that the oil-based products previously contained by the tanks did not have a low flash point.

The small-scale inexperienced contractor was prepared to take on the site as it was in its uncleaned state, and for this reason received the contract. Although previously employed for small jobs by the site owner the contractor had very little knowledge about his capacity or ability to successfully carry out a demolition project as large as Dudgeons Wharf.

The attitude of the demolition contractor, and that of a scrap metal merchant to whom he planned to sell the salvaged metal, was purely a commercial one. Both believed that a good profit could be secured if they worked hard and quickly. An inexperienced and untrained employee of the scrap metal merchant was put in charge of site safety. The site owner similarly appears to have given safety a low priority, although he was aware that the demolition contractor was going to consult the fire brigade. Information on the previous contents of the tanks provided to the contractor by the site owner was vague. The site owner's safety officer assisted the scrap metal merchant's 'safety' person in obtaining some equipment, and in providing a boilerman to operate the boilers that were to produce steam for steam-cleaning the tanks prior to dismantling. Beyond this, the site owner showed no interest in what was going on.

The demolition contractor asked the fire brigade for advice, and the scrap metal merchant's 'safety' employee procured some elementary fire-fighting equipment. However, safety was undoubtedly given a very low priority on the Dudgeons Wharf site. The public inquiry report noted that the demolition contractor:

> ...agreed that he was unaware that any statutory requirements or Regulations applied to the work which was being done and he made no effort, nor did anyone else, to inquire.[2]

While personnel from the brigade's fire prevention section visited the contractor and gave him sound advice, orally and later in writing, they were not experts in dealing with the particular problems faced by the contractor. Because of their lack of expertise in these areas the brigade's fire prevention branch subsequently contacted the acting district factory inspector whom they believed to be experienced in such matters. Unfortunately, it turned out that the acting inspector was less qualified to give advice on the fire safety aspects of the Dudgeons Wharf site than the fire prevention branch. His visit to the site resulted not in a strengthening or tightening up of the safety recommendations but rather a watering down.

However, the public inquiry concluded that if the advice of the brigade's fire prevention branch had been followed by the contractor and, if the contractor had consulted the branch when in doubt, then the disaster might well have been averted.

Within a few days of the contract being entered into, demolition work commenced on the site. On 4 July there was an '8 pump' fire at the tank farm after which the operational branch of the fire brigade expressed considerable concern over the hazards that were present. Due to a misunderstanding – that in future the site owner's safety officer would be supervising the site – the fire brigade's divisional officer relaxed his interest in the site.

The tanks, which were thought to have previously contained turpentine, were steamed in order to render them clean and inert before cutting. However, it was not realised that myrcene, the contents of the tank that later exploded, is an unusual member of the turpentine class of chemicals. Many chemicals leave a deposit on the internal walls of a storage vessel which can be removed by steam cleaning. However, the deposit left by myrcene is not easily removed by steaming. More importantly, the myrcene deposits are combustible, and when ignited or subjected to heating give off inflammable vapour. Apparently neither the site owner nor the demolition contractor were aware of these facts. At no stage prior to commencement of work on the tank that later exploded was the tank's inside inspected or was its internal atmosphere tested with an explosimeter.

The tank was steamed to some degree prior to cutting but, because of the nature of myrcene, this did not remove all of the deposits. Subsequent hot cutting work on the tank ignited the inflammable gas given off by the deposits of myrcene.

As soon as the fire was discovered attempts were made to extinguish it, and the fire brigade was called to the site. However, upon arrival at the scene the fire brigade's officer-in-charge was given incorrect information about the cleaning of the tank and extent of the fire. He was similarly misled as to the vessel's previous contents, the tank having the word 'Turps' chalked on its side instead of 'Myrcene'.

Believing the fire to be out, but deciding to make sure, the officer had some of his men put a spray branch into a manhole on the top of the tank and then ordered that a cover at the bottom of the tank be removed so that it could be visually inspected to see if any fire remained. Since the firemen could not remove the nuts and bolts securing the manhole cover with the spanners available someone suggested that the nuts should be burned off. The cutting torch was applied to one of the nuts and almost instantaneously an explosion took place which blew off the roof of the tank and the six men standing on it. Five firemen and one of the scrap metal merchant's employees lost their lives as a result. The Dudgeons Wharf disaster highlights that even highly trained professionals, who are used to dealing with hazardous situations, can also find themselves to be the victims of misleading information.

7.3 Fairfield Home

On 15 December 1974 at approximately 1.55 am the Fairfield Home for elderly persons at Edwalton, Nottinghamshire was found to be ablaze. The alarm was raised, and the fire and ambulance services were notified of the emergency. Although everyone associated with the incident reacted with all speed, eighteen people died in the tragedy. Seventeen of the fatalities were attributed to inhalation of smoke and toxic gases, and one to a heart attack.

7.3.1 Summary of the incubation period

As with the Coldharbour Hospital refurbishment project, the events leading to this tragedy began with those responsible for providing residential premises for elderly people deciding that the accommodation should be as 'homely' as possible in both its physical appearance and its regulatory regime.

The type of building chosen for the home was originally designed for use as school accommodation. After consultation with both the local fire authority and the fire precaution specialists at the Ministry of Education (because of the design's origin as a school) the county's chief architect believed that the specification and design of the building were suitable in relation to fire. The public inquiry investigators were later of the opinion that the design and specifications of the fire division walls did not in fact comply with the requirements laid down.

Having decided to go ahead and utilise the school-type building the County Council called in an established practice of consultant architects to design the building. The sketch plans were drawn up and sent to the county fire brigade's fire prevention officer for comment. In due course a copy of the plans was returned to the consultants with the fire prevention officer's recommendations.

On the 18 December 1959 the consultant architects applied for bye-law approval from the responsible urban district council (UDC). Unfortunately, however, because of a misreading of section 7 of the Public Health Act, 1936, the council planning officers erroneously believed that Fairfield Home was exempt from bye-law control except for drainage. The public inquiry report notes that this error:

> ...accounts for the building structure not being scrutinised or inspected by West Bridgford UDC during the course of construction, and explains why the advice of the UDC, as the Building Regulatory Authority, was not sought subsequently in 1965 and 1972 when new building regulations came into force, or during the period when the County Architects Department was spending so much time in research into bye-laws and regulations.[3]

(The reference above to research into bye-laws and building regulations in 1972 alludes to the time when the county's architects department realised that the type of building constructed at Fairfield could be vulnerable to fires but because no regulation appeared to exist to deal with the problem they were confused as to what action to take. In order to reduce their anxiety they carried out a comprehensive survey to ensure that all their building stock conformed to the current building regulations.)

Work commenced on the Fairfield Home building in April 1960 and the first residents moved into the completed accommodation in November 1961. In 1966 the consultant architect visited the home to see how well the building was performing. Apart from a few minor complaints regarding soundproofing of the staff's quarters everyone appeared to be happy with the building.

From the time the home was opened a number of problems manifested themselves and attempts were made to provide solutions. For example, because of the need to comply with bye-law 100, which in essence states that a water closet cannot face on to a room intended principally for human habitation, fire doors had to be located between the central lounge area, bedrooms and the water closet. Every time a resident wished to use the toilet it was necessary for him or her to pass through the self-closing fire doors. The springs providing automatic door closing were rather strong, and old or frail residents had a great deal of difficulty in negotiating them. As a consequence it became a practice of the residents to wedge the doors open.

The problem of residents wedging fire doors open was noted by the local fire prevention officer on his inspections of the premises as early as 1968. He passed this information on to the chief fire officer who, in reporting his observations to the council's welfare officer, suggested that the installation of auto-close doors operated by smoke detectors would solve the problem. The chief fire officer also asked that the council consider installation of smoke detectors in all elderly persons' homes of greater than one storey in height to enable the earliest possible warning of fire to residents and staff.

While money was eventually made available for the detectors the programme priority was very low for single storey homes, and by the time of the fire no smoke detectors had been installed at Fairfield. Also by the time of the fire the hazard of unstopped roof voids, knowledge of which had led to the 1972 survey of buildings, was generally well understood and remedial work was being carried out in premises under the county council's responsibility. However, given government expenditure cut-backs at the time, along with other demands, the county's chief architect took the decision to give priority to schools. Little danger was perceived in homes like Fairfield and there was no sense of urgency to carry out remedial work.

By June 1973 the county council's elderly persons' homes had been surveyed by the fire brigade's fire prevention officer to ensure that they complied with building regulations. In November 1973, Fairfield was upgraded to provide a special degree of care where the majority of residents required exceptional caring skills. This created a situation which had not been envisaged when the home was originally designed.

The chief fire officer had by this time given his permission for the fire doors to remain open during the day. But the self-closing mechanism still caused the disabled residents problems. Consequently, some residents continued to wedge open the fire doors at night.

The original concept behind the accommodation at Fairfield was that of 'homeliness'. The residents were to be allowed as far as possible to retain their dignity and preserve privacy while having an opportunity for interaction with others. In this environment residents were allowed to smoke in the general areas but were warned of the dangers of smoking in their bedrooms.

Because the home's residents were elderly people the central heating system was set to maintain a temperature of 75°F. This tended to dry out the furnishings and the timber of the building, making the home more vulnerable to fire.

During the night of 14/15 December one of the home's residents dropped a cigarette or a lighter in her bedroom and the carpet began to smoulder. The fire was soon discovered by a resident who went off to find a member of staff. However, the member of staff on duty had been called away to another part of the building. She had needed assistance and had gone to wake up the deputy matron for help. By the time the resident found the staff members the fire had established a firm hold.

Immediately the matron and her assistant raised the alarm, the fire brigade was informed of the emergency and efforts were made to get the residents out of danger. Within a short period of time the fire brigade arrived and started to extinguish the fire and assist with the evacuation of residents.

The evacuation was difficult since all the residents were frail and could only move slowly. A number had taken sedatives, while some had profound physical and mental handicaps. In addition the ground floor windows were difficult for people to get through since they had not been designed as fire exits.

This disaster in particular highlights how a hazardous situation can arise through a number of incremental changes occurring within an organisation. While each individual change is small the sum of those changes can be extensive. The result is that the organisation can, as in this case, find itself operating in a manner which had not been envisaged. As a consequence the difficulties which have to be overcome by organisational personnel in an emergency are far greater than those planned for.

7.4 Summerland

On the evening of 2 August 1973, the Summerland leisure centre complex in Douglas, Isle of Man, was inadvertently set on fire while containing approximately 3,000 holiday-makers and 200 staff. The building, an open-plan structure whose external envelope comprised steel sheets and transparent acrylic panels, burned rapidly. Fifty people lost their lives.

7.4.1 Summary of the incubation period

The economy of the Isle of Man is heavily dependent upon income generated during the tourist season. During the 1960s the trend began for British holiday-makers to take their holidays abroad. The islanders responded by looking at ways to attract the tourists back. The original concept of Summerland was to recreate within the building the atmosphere and activities found in a Cornish seaside village, but with the advantages of an artificial Mediterranean climate. There were to be places to shop, drink, sunbathe, swim and listen to music even on the worst days of the island's summer.

The first phase of the project called for the construction of a swimming pool called the Aquadrome. This was carried out and the pool opened in 1969. During the design stage of phase two the original innovative concept of a village changed into something more practical, accommodating music venues, licensed bars, amusement arcades, etc. The eventual Summerland building was designed to abut the Aquadrome. The internal structure of the centre was a modern open-plan style on several floors, some arranged in the form of terraces overlooking the central entertainment area.

The leisure centre was developed by Douglas Corporation, the local authority of the largest town on the island, with financial assistance from the government of the Isle of Man. The shell of the completed building, owned by Douglas Corporation, was leased to a leisure company which had the authority to design and build the more decorative parts of the interior. There was, however, an important gap in the continuity of the project between the design and construction of the shell under one design team, and the design and furnishing of the building by a second design team employed by the lessees. In addition to its role as developer, the local authority was involved through its planning, engineering and fire safety committees in scrutinising successive bye-law, planning and safety submissions with regard to the building as design and construction progressed.

The design of the shell of the building was placed in the hands of a local architectural practice. Because the practice had never been involved in a project of this size before it obtained agreement for a larger practice on the UK mainland to be retained as associate architects. In the second phase, the design and fitting out of the interior, the mainland practice was retained by the lessees of the building as principal architects.

At approximately 7.40 pm on 2 August 1973 three schoolboys who were on holiday at the resort inadvertently set fire to the remains of a fibreglass kiosk which had been used earlier in the summer as a ticket booth for a miniature golf course. The kiosk had been damaged in a storm and, although a large part of the booth structure had been removed from the terrace, one section had been left stacked close to one of the walls of the leisure centre.

The fire once ignited swiftly took hold and soon the remnants of the kiosk were burning fiercely with the flames coming into contact with the wall of the leisure centre, which at this point was constructed of corrugated sheet steel coated on both sides and sold under the trade name of Colour Galbestos. The heat generated by the fire produced vapours from the coating on the inside of steel sheet which also ignited, thus introducing fire into an internal void which, through poor control procedures during the building's outfitting, had been lined with materials which were not as fire resistant as they ought to have been. As a consequence, the fire rapidly swept through the void and subsequently by other routes, starting at about 8 pm, to other levels of the leisure centre.

The south wall and the roof of the building were glazed with Oroglas sheets, an acrylic material which, while it had been utilised elsewhere with success to glaze other types of 'space structures', had never been employed on the same scale as at Summerland. As the fire spread to the south wall and roof and began to consume the Oroglas it rapidly developed into a conflagration of massive proportions destroying almost everything which would burn on the top four levels of the complex. The roof was totally consumed by the flames in about ten minutes.

Members of the Summerland staff, using fire extinguishers and a hose pipe, made unsuccessful attempts to put the fire out while it was still in its early stages. The automatic fire alarm system was overlooked, and the house manager called the fire brigade at 8.01 pm from a public telephone box. Members of staff eventually operated the internal fire alarm at 8.05 pm. By this time, however, the fire brigade had already received two reports of the conflagration, one from a ship out at sea, and was on its way. Regrettably, the alarm when operated only sounded in the local fire station and not in the leisure complex. Thus the patrons in the building remained unaware of the life-threatening situation that was developing.

Within the building, uncoordinated attempts were made to evacuate some patrons, and to avoid creating panic amongst others. Unfortunately, the house manager, when switching off the main electricity supply to the building at 8.11 pm, did not notice that the emergency lighting generator had failed to start. As a consequence many of the centre's internal staircases were thrown into darkness thereby hampering patrons in their attempts to evacuate the building or find their children.

Compounding these problems was the fact that some emergency doors were found to be locked, and time was lost as attempts were made to either break open the doors or find duplicate keys. Other emergency doors were found to be blocked from the outside by vehicles that had been illegally parked. This created serious difficulties both for those trying to evacuate the building and for the fire-fighters attempting to bring the blaze under control.

Smoke and flames rapidly swept through the building as a number of the patrons who were still inside tried to find a way out or locate their children in other parts of the complex. Many people fell over in what became a scramble to escape the flames. Although no deaths resulted from this a number of people were injured. Within a very short space of time from the fire breaking out of the internal void and into the main entertainment area those who had not successfully managed to make their way to a clear exit were dead.

The subsequent public inquiry paid attention to various factors, including: the extent to which informal contacts between those in the island community were developed at the expense of more formal procedures; the pressure for completion of the second phase of the building in time for the tourist season; significant changes in the assumed operating conditions of the building as the design progressed; some very ill-defined and poor communications; the issue of waivers to building regulations; the inappropriateness of the fire regulations then in force to the Summerland project; the organisation of staff at the leisure centre – in particular the apparent lack of arrangements for the training of staff in fire procedures.

The Summerland disaster demonstrates how political and economic considerations can fuse with other types of organisational pathologies to create a complex web of circumstances which can lead to an unperceived hazardous situation existing. The disaster had far reaching effects not only for building regulations on the Isle of Man but also for the UK as a whole.

7.5 Taunton railway fire

Sometime during the early hours of 6 July 1978 a number of sacks of soiled and fresh linen in a sleeping-car on the overnight Penzance to Paddington train started smouldering and then burst into flames. The sacks had been placed against an electric heater in the vestibule of the leading sleeping-car. Because of the flammability of the furnishings and fittings used in the coach the fire was soon well established.

Some time after leaving Exeter, the sleeping-car attendant left his coach to have a wash and brush up in berths 1 and 2 of coach C. As he left the lavatory he smelt burning and found smoke coming from his coach. He tried to raise the alarm and to go forward but was driven back by the smoke and shortly afterwards was rendered unconscious. The sleeping-car attendant in the next coach also smelt smoke and pulled the communication cord to raise the alarm.

The train, travelling slowly at that time, quickly came to a stop. The signalman, contacted by the driver of the train via a nearby track side telephone, then arranged for the emergency services to be called. Although everyone concerned acted with commendable speed, eleven people lost their lives, either directly or indirectly through the inhalation of smoke, while another sixteen sustained various types of injury.

7.5.1 Summary of the incubation period

One of the main factors underlying this disaster was failure by the relevant authorities to perceive that a sleeping-car coach, whilst clearly being part of a train, is also effectively a small mobile hotel. Provision of fire precautions, fire-fighting equipment and passenger evacuation methods need consideration to a similar, if not greater, extent than for normal hotel premises. A sleeping-car is frequently moving at speed, which can hamper both evacuation and fire-fighting. At the time of the tragedy, sleeping-cars were not required to have formal inspections with regard to such matters, unlike hotels and offices.

The sleeping-cars used on the night of the fire were originally constructed in 1960, with a steam heating system provided by the hauling locomotive. During 1977, this system was supplemented with an electric heating system. Of particular importance was a modification providing an electric heater in the sleeping-car attendant's pantry and another heater on the wall of the vestibule outside. These were wired so that one or other of the heaters was always on.

Another important factor leading up to the fire occurred in May 1977 when the brake-van, where the sacks of linen had been stored, was removed from the formation of the train.

After this procedural change the linen was kept in the vestibule in the sleeping car, contrary to regulations.

The design of the ventilating system also had a part to play in the tragedy. The internal source of cold air supply for each compartment came from an intake adjacent to the source of the fire. Smoke was drawn through the ventilation system and into each sleeping compartment with open ventilation louvers. The smoke also travelled along the corridor and entered the sleeping berths via ventilators set in the bottom half of the compartment doors. According to the inquiry report, most of the dead were:

...overcome by fumes without fully waking.[4]

Training of the sleeping-car attendants left much to be desired. The attendant responsible on the night of the fire held an erroneous belief about the requirements for locking carriage doors. He also had never seen, or been given, a copy of the regulations relating to his post.

One possible means of evacuation was the drop-light window in each berth. Many of the passengers were not aware of this escape route. Only one passenger involved in the fire said that he knew prior to the disaster of the existence of the drop window exits. An additional problem was that the call-bell in the coach had been deliberately disabled so that there was no ready means to call the attendant.

In this case study it can be observed, as with some of the other disasters, that prior to the event the hazardous situation had not been perceived by organisational personnel. The public inquiry report noted that the special nature of fires in trains, and in particular sleeping cars, had not been fully appreciated. In the aftermath of the accident it was deemed that the regulations applying to hotels should also be applied to sleeping-cars as far as possible. Eventually the recommendations of the report were also incorporated into the design of a new class of sleeping-car. Provisions were also made for an improvement in training and supervision, communications, and operational procedures.

While the case study descriptions above are simplified thumbnail sketches of the actual events and analysis which took place following each incident, they do demonstrate some of the complexity and interrelation of events that can lead to disasters. In each case it can be seen that the different actors involved were not fully aware of the hazards they faced, and were consequently both surprised by, and unprepared for, the events which unfolded before them. It is the lessons that have been gained from these and other painful events which have been synthesised and incorporated into the models presented in this book.

Notes

1. *Coldharbour report*, op. cit.

2. *Public inquiry into a fire at Dudgeons Wharf, on 17th July 1969* (1970) Cmnd 4470, HMSO, London.

3. *Fairfield report*, op. cit.

4. *Taunton sleeping-car fire report.*

Chapter 8

Discussion and conclusions

8. Introduction

This final chapter reviews some fundamental themes, before summarising and reviewing the research upon which the book is based. The summary includes observations and conclusions regarding the field investigations, interviews and documentary evidence, and provides an overview of the theoretical framework developed. The bulk of the chapter discusses a proposed systems-based organisational model that, if adopted, could be a basis for a national system to reduce socio-technical failures. The model consists of a hierarchy of subsystems, some of which would feed back information on socio-technical failures through several paths to industry, designers, legislators and educational establishments, to aid organisational learning.

8.1 Fundamental themes

Several themes underlie the notions discussed in previous chapters, and it is worth summarising them here:

- Large-scale accidents – disasters – are generally failures of socio-technical systems.

- Typically, disasters are low frequency events within any single industry or field of activity.

- Each industry or profession needs to look outside its own immediate scope, and examine disasters occurring elsewhere, in order to maximise the chances of finding repeated patterns and of learning isomorphically.

- The goal of the isomorphic learning process is 'active foresight' regarding safety. Both elements of this goal are equally important – the generation of foresight, and the active implementation of risk reduction procedures and systems based on the knowledge gained.

- An organisation's safety culture – the sum of beliefs, norms and attitudes regarding risk and safety – determines the extent to which foresight is generated, and the level of implementation.

8.2 Review of the research

8.2.1 Are public inquiries appropriate?

At every large-scale accident inquiry the hope is expressed that the investigation will ensure that 'this shall not happen again'. In practice, adequate learning is often constrained. Frequently there is an assumption that the particular large-scale accident is unique and unlikely to recur – even, perhaps, that it is pointless to look for regularities in such 'acts of God'. Often there is also no readily available perspective to interpret findings at an appropriately general level. More fundamentally, there may be no set of techniques for discerning all the relevant patterns in the events surrounding an accident.

Some of the people closely involved with public inquiries as attendees, interviewed during the course of this research, have argued that such inquiries are not always the formalised, objective, truth searching bodies of the common public perception. Public inquiries have no laid down formal procedures, are adversarial in nature, have no power to require organisations or individuals to carry out their recommendations, and may sometimes apparently have hidden political agendas to address.

Support for this view comes from Levitt and March who argue that:

> ...the organisational, political, and legal significance of [low probability, high consequence] events, if they occur, often muddies the making of inferences about them with conflict over formal responsibility, accountability and liability.[1]

While all of those interviewed for this research believed that some form of public investigation into major accidents was required, many of the respondents expressed a desire to see the current system change to one where blame and political expediency were not at issue. The expression of such sentiments casts doubt on whether the current structures and methods of public inquiries are appropriate for investigating, analysing, reporting, recommending and disseminating information regarding catastrophic events.

8.2.2 Better utilisation of inquiry information

However, it is not likely that public inquiries will radically change their form in the near future. They are likely to remain the most prominent form of disaster inquiry, and it is vital that the best use is made of their investigations and recommendations. In this research, analysis of recommendations made by selected public inquiries revealed that those recommendations could be placed into a number of categories and sub-categories at differing levels of analysis. What are the reasons for similar recommendation types recurring in investigations into different types of disasters?

Three questions might be asked about such similarities or isomorphisms. Do they occur because people investigate events in a limited number of ways? Do they occur because organisations can only learn from their failures in a limited set of ways? Or, do they occur because disasters have similar aetiologies? It is argued here that the last of these reasons is the more likely. If this is correct, all engaged in investigations, and in the design and management of socio-technical systems, owe it to those who have been involved in disasters to learn as much as possible from those events. More research is required to help formulate a framework which will aid industry, commerce, the professions and government to learn from, and avoid, disasters.

8.2.3 Summary

Several conclusions can be derived from the research. First, disasters are socio-technical in nature as opposed to solely, or even mainly, technical. This is reflected in the public inquiry recommendations: those relating to 'social' factors are always greater in number than those which refer to the technical aspects of a disaster.

Second, the evidence strongly suggests that it is usually failures in organisational arrangements which ultimately lead to technical failure. In many ways this is not surprising since it is people who conceptualise, design, develop, construct, operate and maintain both the physical and organisational systems which sometimes fail so catastrophically.

Third, public inquiry recommendations sometimes do not address all of the chains of events leading to the disasters under investigation. This implies that the recommendations, even if implemented in full, will fail to cut some of those chains if they recur in the future in the same or in isomorphic systems. Development of the schematic report analysis diagram (SRAD) method of analysing inquiry reports would be a step towards resolving this problem. The computerisation of this technique could be especially useful in disaster analysis and reporting, both internal and external to an inquiry, and also as a teaching aid.

Fourth, analysis of interview data and documentary evidence showed that organisational responses to disasters could be placed into two broad categories. In one category were responses relating to general organisational and social concerns. In the other category were the responses relating to specific areas of organisational life. Both categories could be further divided into sub-categories, the subdivisions being derived from the data. While interviewees, and other sources, often did not explicitly refer to many of the subcategories, their replies highlighted that they implicitly utilised, and continued to utilise, many of the notions when making operational decisions.

8.2.4 Models of active learning

Using the two categories of responses, and the subcategories within each of them, two models were developed to describe how organisations endeavour to learn from catastrophic events. The general and specific categories of organisational concerns form the 'steps to active foresight' and the 'steps to active learning' models respectively. Together, these models constitute elements that can influence an organisation's 'sentient system' or its safety culture.[2] This is one way of regarding the system of beliefs which creates the environment in which organisational learning can take place.

Thus, for example, in the aftermath of a disaster, should the organisational personnel involved (or those in other organisations who might learn isomorphically) not be sufficiently 'surprised' at what has taken place, not be emotionally affected by the event, or believe that the incident was an 'act of God', then learning from the catastrophic event is less likely to take place and the stage could then be set for a recurrence.

If an organisation's safety culture is so defective, such factors will have a negative influence on the attainment of foresight. It might also be the case that the 'safety by compulsion' category is also flawed, and no legislation is enacted aimed at preventing recurrence of similar behaviours/system conditions to those which pertained prior to an accident. In these circumstances either or both of the inputs to the 'steps to active learning' model might be inoperative, and as a consequence no active learning would take place. Conversely, a positive safety culture and proper legal framework would give appropriate operative inputs to the 'steps to active learning' model. The internal organisational culture and the external regulatory environment would both be positive, and hence inputs to the 'steps to active learning' model would be fully activated.

The elements of the 'steps to active learning' model also influence an organisation's safety culture (Figure 6.1). If the inputs to the 'perception of stability unit' are defective then appropriate error signals will not be generated. This might be the case if there is any deficiency in the inputs regarding information and disasters, preconditions to disaster, safety by regulation, or lessons learned about fires. On receipt of error signals, the level of any control action from the 'stabilising action unit' will be determined by the staff who have to implement active learning measures, and by the impact of organisational economics.

These staff and economic factors are of great relevance. Even though senior management may wish to implement all of the measures recommended by a public inquiry, it may be the case that other organisational personnel do not wholeheartedly embrace the necessary changes, and active learning will be limited. While this is less of a problem in organisations which have been involved in a disaster, it might be a major factor where one organisation is attempting to learn isomorphically from other organisations. This can give rise to failures of both foresight and hindsight. Thus, it is of paramount importance that organisations should develop a 'safety imagination' and acquire a safety culture in which personnel positively affect active learning.[3]

A striking similarity is observed between the issues addressed in public inquiry recommendations and the specific types of organisational concerns noted by respondents in this research. However, one category – 'organisational economics' – is almost solely to be found among the concerns of the organisations and personnel involved. Typically, no reference is made in a public inquiry report to the costs which might be incurred by an organisation which engaged in the active learning recommended by the inquiry. While cost often does not appear to be seen as a problem for those organisations involved in a disaster, it may be of central concern to organisations which may not have been involved but which wish to learn isomorphically from the event.

Evidence presented in Chapter 6 in relation to organisational 'trade-offs' suggests that this is indeed the case. Thus, one way to encourage organisations to learn actively from the failures of others might be for a public inquiry to cost their solutions so that they derive the most economical recommendations. Another possible answer, depending on the circumstances, would be the provision of central government assistance to defray some of the expenses incurred by organisations in implementing inquiry recommendations. However, given the preoccupation of many governments with cutting the public sector, it is highly unlikely that such funds would be forthcoming regardless of merit.

8.3 Model for the future

Perhaps one of the most crucial elements in the development of adequate organisational learning is that of finding appropriate mechanisms for the generation and dissemination of information gained from unwanted events. The accumulation of such information is desirable. Roberts and Burwell note in their study of learning within the nuclear power industry:

> In any system, one of whose desirable attributes is safe operation, the rise of cumulative experience will tend to result in a reduction of accidents per unit operation.[4]

In a similar vein, Levitt and March have observed that there is strong evidence to suggest that direct cumulative experience results in positive outcomes for organisations, in that unit costs go down as experience increases. The equation derived from such empirical results is:

> ...similar in spirit and form to learning curves in individuals and animals... Much of the early research involved only simple graphical techniques, but more elaborate analyses have largely confirmed the original results.[5]

Therefore, if the direct cumulative experience of those organisations involved in disasters can be placed at the disposal of other organisations who could then benefit from such experiences, there is a strong possibility of reducing the risk of disasters. There are, however, many problems surrounding the dissemination and use of information stemming from socio-technical failure. Not least is the fact that currently no system exists at national level for co-ordinating and synthesising the information generated by catastrophic events.

Various bodies do good work in disseminating information in their various fields. An example is the *Loss Prevention Bulletin*, a periodical comprising case studies of disasters and near-misses, published by the Institute of Chemical Engineers (IChemE). Other examples include the various mandatory and voluntary accident and incident reporting schemes in the aviation community. Bodies such as the Royal Aeronautical Society (RAeS) and the Nautical Institute have held conferences that contrast and compare safety in the air and at sea, largely with reference to past disasters. These are valuable attempts to aid learning. However, there is no national organisation that provides overall co-ordination for all such initiatives.

While the UK's Health and Safety Executive (HSE) does have wide responsibilities regarding health and safety at work, including the collection and dissemination of accident information, its scope does not cover all industries. For example, medical or aviation disasters are excluded from the HSE's ambit. There is no standardised way of communicating knowledge, gained so painfully, to organisations who might require it. Each investigating body has its own thoughts on how, and to which organisations, its information should be disseminated. This fragmentary situation makes it difficult to perceive the isomorphic lessons which might be present in the data and might further aid our understanding of such events.

8.3.1 Model of an organisational learning system

One way forward might be to have an organisation or cluster of organisations, at a national level, dedicated to the collection, analysis and the dissemination of information gained from all kinds of socio-technical failures. Like any other organisation or system, such a proposed body should not be thought of as a single level structure, but as incorporating linkages between a number of levels and a variety of organisational learning process. Figure 8.1 is one way of illustrating such a structure.[6]

The model illustrated is an attempt to create a theoretical 'system of systems' which might help to reduce the number of socio-technical failures. It is sketched out here not as an immediate policy proposal, but in order to highlight those issues which any such system would need to confront. In the model the terms 'environment', 'designer' and 'manager' should be interpreted very broadly, so as to include all those situations and personnel involved in the development and operation of a socio-technical product, organisation or procedure.

Referring to Figure 8.1, a sequence of events would typically begin with someone perceiving a need for change in the 'environment', such as the need for a new bridge, a new power station or new hospital ward. Once the need has been recognised, and sufficient financial backing made available to support the project, the next event would be specification of the precise attributes of the finished product, organisation or procedure. This would occur within the 'design specification system'. For example, the specification might be that a bridge should be capable of withstanding 'x tonnes' load, or that a power station should be capable of producing 'y megawatts' power output.

Having decided upon its parameters, the design specification would now be submitted to the individuals and organisations contributing to the 'design system'. Here the specification would be translated into a series of working plans representing the finished product. After the design process is completed, and the design is accepted by the client, it would then be transferred to the 'design implementation system'. At this stage the design is costed, contracts and subcontracts issued, and the construction or building work is undertaken. When the implementation process has been completed, operational instructions are issued. The 'operational socio-technical system' so created then begins to act, and in so doing changes the environment in the manner originally stipulated in the design.

What if the socio-technical system now fails – say, a bridge collapses – or the change in the environment creates unforeseen problems for the other organisms sharing that environment? Among the elements of the model described so far, there is no mechanism for the lessons from such experiences to reach all those sections of society which might usefully gain from such information.

The right hand loop in Figure 8.1 is such a mechanism – a 'design learning system' – that would help the lessons drawn from socio-technical failures to be incorporated into the working practices of the future. To achieve this, the 'design learning system' would need to collect, collate and analyse data from known socio-technical failures, and also to incorporate knowledge gained in the sciences, fusing together both practical experience and academic research. The output of such a system would be valuable knowledge for the design and management functions.

This knowledge would be particularly valuable if, as the model indicates, it could then be made to impact upon on the designers of socio-technical systems, and the legislators and managers who control such systems. Beyond this, information relevant to the management of socio-technical organisations could be communicated directly to the industries concerned, and also to educational establishments and professional institutions, and others. The whole system of systems could perhaps be managed and monitored through a government body or multi-professional institution.

8.3.2 Expanded model

An expanded version of the model is shown in Figure 8.2. This illustrates how a complex system can be constructed from less elaborate subsystems. It also demonstrates that relationships between the subsystems need to be organised in a coherent way. While there would be numerous problems involved in creating and monitoring such a socio-technical failure reducing system, it would be more manageable if broken down into constituent parts. One of the central tools required for such a task has been forged through the development of the 'systems approach' to complex problem solving.[7]

In the expanded model, three levels of organisation are illustrated. The first level contains the 'design implementation system' and the 'operational socio-technical systems'. The second level contains the first level, plus the 'design process' and the 'formulate design specification' systems. The third level contains the first two levels, as well as the 'design learning system'. The complexity of the overall system grows as each new level of organisation is added.

Figure 8.2 shows a sample of the possible and necessary interconnected subsystems. The links highlight the importance of communication and co-operation between subsystems if optimum solutions are to be arrived at. A sequence of events can be followed through Figure 8.2 in a similar manner to that already discussed for Figure 8.1. A desire for a change in the 'environment' prompts the formulation of a 'design specification' for the project, which will ideally be devised with the benefit of opinions sought from all 'concerned actors' in the system. The completed specification will then be translated into firm proposals in the 'design process' system, aided by activities of the 'problem solving' and 'collation of designs' subsystems. The latter co-ordinates the designs from the various separate sections of the project to prevent conflicts.

A 'simulated systems' subsystem is included in the model here to emphasise the importance of non-destructive testing of proposals in as many failure modes as possible, before the design is finalised. Here, as in other parts of the system, good communications and a two-way dialogue are important in reducing the possibility of failure.

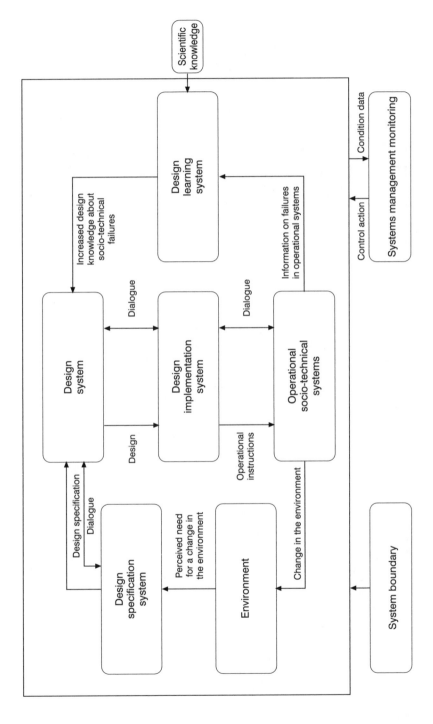

Figure 8.1 A socio-technical failure-minimising system

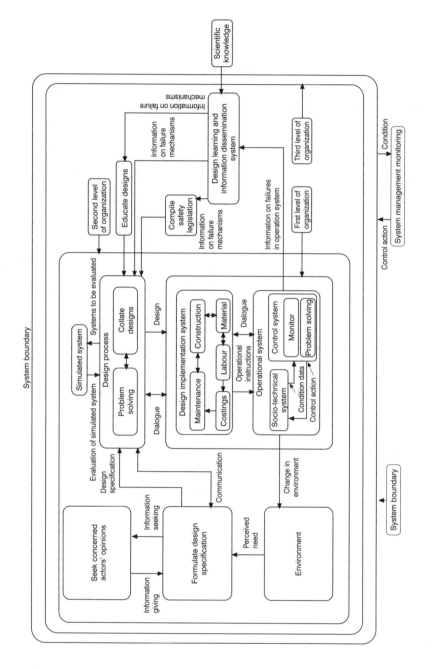

Figure 8.2 Expanded socio-technical failure-minimising systems model

As in the simplified model, the 'design implementation system' translates the design drawings into 'operational instructions' which include parameters relating to cost, labour, materials, maintenance, construction, etc. Just a token sample of possible subsystems have been selected here for illustrative purposes.

Once operational instructions have been issued the 'operational system' can be brought into action, setting up the required 'socio-technical system'. This 'operational system' should generate both the environment changes initially specified for the project, and also 'condition data' that flows into a 'control system'. This 'control system' loop is utilised in monitoring the operation, and to initiate corrective control action where possible. This process is directly analogous to the negative feedback loop described in the 'systems failure and cultural readjustment model' of Chapter 2.

Within the 'design learning system' only three subsystem paths have been highlighted. This is an adequate number to express the general notions involved. The 'design learning system' would receive information from various sources on socio-technical system failures, and also from academic bodies. The data would then be assimilated within the 'socio-technical failure accumulation and promulgation of information process' subsystem. Once the data had been analysed to extract useful information, the subsystem would transmit that information about failure mechanisms:

- to 'designer educating' and 'manager educating' subsystems;

- to subsystems concerned with the production of codes of practice and safety legislation;

- directly to industry via trade journals, special notices (as occurs in civil aviation), pamphlets, conferences and specially organised courses.

These information flows would produce designers and managers who were much more aware of the problems and mechanisms of failure. They would aid the drafting of codes of practice and safety legislation, and so would impact upon future work procedures and design practices. The information flows would bring industry up to date faster than currently regarding erroneous practices and dangerous conditions.

The 'design learning system' can be thought of as acting as an explicit negative information feedback device in a closed loop system. The information which it would analyse comes from the output of a dysfunctional system, and is fed back in the opposing sense to that dysfunction so that the deviation monitored is reduced or does not recur.

8.3.3 Model discussion

While the model described above does have a number of drawbacks, not least that such a system would be difficult to devise and operate, it does provide a framework for the discussion of organisational learning from disasters. It might help to change the way in which industry, commerce and regulatory authorities view the utility of information derived from such sources. As Turner and Toft note:

> The difficulties in moving current practitioners a very small step along the way towards the constructive use of failure-derived information indicates clearly that a sociological analysis of current practices in any industry operating large-scale hazardous systems would reveal a pattern much different from the model outlined, with the partial exception, perhaps, of the nuclear industry.[8]

It is argued here that if a system such as that suggested in the above models were to be realised, it would not only aid our understanding of catastrophic events but also legitimise socio-technical failures as a field of study. The creation of this field of enquiry might well bring about a positive change in the way both individuals and organisations perceive and react to such failures, and hence help to create a 'climate' or culture in which safety related matters could be discussed more openly.

8.4 Conclusions

There is little doubt that active learning can assist organisations. Levitt and March note:

> The speculation that learning can improve the performance, and thus the intelligence, of organisations is confirmed by numerous studies of learning by doing, by case observations, and by theoretical analysis.[9]

However, from the evidence presented here it appears that while many organisations and their personnel do actively seek to reduce the possibility of a disaster, there are others who hold the view that such events cannot be prevented and hence that learning is either impossible or a waste of time. There are always those who believe nothing can happen to them. It also appears that while 'time is a great healer', it can equally sow the seeds of destruction. As time passes an organisation tends to forget the lessons of history, as personnel, routines and procedures change.

Discussing why warnings of disaster are sometimes ignored, Turner points to the work of Martha Wolfenstein, who suggests that one of the reasons for such behaviour is:

> ...the sense of personal invulnerability which is essential for most individuals to maintain if they are to be able to go about their daily business without constantly worrying about all the possible dangers that could threaten them.[10]

Economic and political factors are of particular significance in determining the amount and type of learning from major accidents. When all or several of these powerful human and organisational factors exist in an organisation it is hardly surprising that failures of both foresight and hindsight occur, or that active learning consequently fails on a permanent basis. However, as with many organisational pathologies, once such problems are recognised it is often possible to develop strategies to overcome them providing there is a will to do so.

Finally, it is clear that while many organisations can and do learn from disasters, others do not. Some of the pathologies which inhibit learning, and which affect the ways in which organisations try to learn from major accidents, have been identified and discussed in this book. Evidence suggests that disasters are not beyond our comprehension and consequently we can, to some extent, control the risk of their occurrence. While perfect safety is always likely to be a utopian vision in the context of commercial and industrial organisations, it is the authors' contention that society should attempt to approach that ideal as closely as possible.

Notes

1 Levitt and March, op.cit.

2 ATOM (1987).

3 Pidgeon, N.F. (1988) Risk assessement and accident analysis, *Acta Psychologica*, 68, pp 355-368.

4 Roberts, P.C. and Burwell, C.C. (1981) The learning function in nuclear reactor operations and its implications for siting policy, Oakridge Associated Universities, May.

5 Levitt and March, op. cit.

6 Toft, B. (1984) *Human factor failure in complex systems*, Dissertation, for the awarding of a BA (Hons) degree in Independent Studies, University of Lancaster.

7 Checkland, op. cit.

8 Turner and Toft, op. cit

9 Levitt and March, op. cit.

10 Turner, op. cit.

Index

Crime at Work: *studies in security & crime prevention*

Topics covered include:

robbery, commercial burglary, ram raiding, shoplifting, insurance fraud, violence against staff, crime on industrial estates, cheating in hotel bars, terrorism and the retail sector, the effectiveness of electronic article surveillance, customer and staff perceptions of closed circuit television, security implementation in a computer environment, and the advantages of in-house to contract security staff.

Edited by Martin Gill

This groundbreaking book contains a wealth of information which will be essential reading for all those interested in crime prevention, security, the motivation of different types of offenders, and the effectiveness of various security measures. Each article covers the theme of crime prevention. Papers incorporate the views of offenders, victims, customers and staff.

Until now there has been very little consideration of the extent, impact and patterns of crimes that occur in the workplace. This important text suggests that such an omission is no longer justified. Produced in collaboration with business, the book reflects the growing realisation that effective responses to crime are based on the need to collect and share information.

Crime at Work: studies in security & crime prevention

"This book breaks new ground in many areas and contains a wealth of interesting facts and hard information."
Commercial Crime International

£25.00
ISBN 1 899287 01 9
240 pages
(index included)

1994

Issues in Maritime Crime: *mayhem at sea*

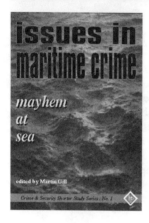

Crime and Security Shorter Studies Series: No. 1

Topics covered include:

fraud, piracy, arson, theft, deception, smuggling and drug trafficking. There is also a focus on containerisation, boat watch, insurance, registration and marking schemes, physical security measures and their potential to prevent offending.

Edited by Martin Gill

This path-finding book offers new insights into aspects of maritime crime and its prevention. Its coverage of both domestic and international issues will appeal to all those interested in crime prevention, security and maritime issues.

Articles have been written by internationally recognised experts on maritime crime. This includes the police, HM Customs and Excise, a private investigator, as well as independent specialists and academic researchers.

Issues in Maritime Crime: mayhem at sea

Papers refer to real cases offering a fascinating insight into the threat posed by crimes that occur at sea. The information in this text suggests that internationally and domestically the official response to maritime crime has too often been unimaginative, misdirected and partial and has sometimes worked against the interests of crime prevention.

"Any owner who is realistic enough to appreciate that crime is not merely something that happens to other people, would be well advised to study these papers."

£12.95
ISBN 1 899287 02 7
80 pages
(index included)

Practical Boat Owner

1995

Crime and Security Shorter Studies Series: No. 2

Topics covered include:

theories of crowds, police policy and training practices, developments in strategy and tactics, the policing of political, industrial, festival and urban public order events, the future of policing in a 'post-modern' society.

Mike King and Nigel Brearley

This highly commended book highlights the major 'watersheds' in the policing of political, industrial, festival and urban disorders and contains a wealth of material from interviews with senior police officers. The book is written in a clear and concise style, incorporating an extremely informative glossary of terms. This work will be essential reading for both police practitioners and those studying or interested in the area of contemporary policing.

Public Order Policing: contemporary perspectives on strategy and tactics

"Mike King and Nigel Brearley are to be congratulated on this well documented and compelling analysis of the changing face of British public order policing. This timely and refreshingly accessible book ought to be essential reading for anyone seeking to understand the recent evolution of police crowd control methods, or eager to predict the future direction of police public order strategy and tactics."

£14.95
ISBN 1 899287 03 5
128 pages
(index included)

Dr David Waddington
Sheffield Hallam University.

1996

International Journal of Risk, Security and Crime Prevention

International Journal of Risk Security and Crime Prevention

VOLUME ONE Number One

January 1996

Published by Perpetuity Press

The Journal offers new ideas and insights in the area of risk, security and crime prevention. Those who subscribe to the Journal will be able to learn from the findings of independent research which shows how to identify new opportunities and make informed decisions. The Journal boasts a prestigious advisory board. All papers are written by experts and are independently refereed.

Articles include: business crime, security shutters, risk and town centres, crime and loss control training, deviant drivers and moral hazards, stalking, shop theft, organised crime, robbery, commercial burglary, household property crime, crimes against the business sector, crime and unemployment, evaluating the effectiveness of CCTV, security by design, private security regulation, prior victimisation and crime.

The Advisory Board

The Editorial Advisory Board includes nominees from: the American Society for Industrial Security, the Association of British Insurers, the Association of Security Consultants, the British Bankers' Association, the British Retail Consortium, the British Security Industry Association, the Building Societies Association, the Health and Safety Executive, the Institute of Risk Management, the International Professional Security Association, the Loss Prevention Council, Victim Support and others.

From January 1996 Annual Subscription Rate (1996): £130.00 (£145.00 Overseas)

Single issues £40.00 (£45.00 Overseas) ISSN 1359-1886 280x210mm

Quarterly

International experts include Professor Joshua Bamfield, Professor John Benyon, John Burrows, Professor David Canter, Dr Marc Cools, Bruce George MP, Professor Mike Levi, Dr Rob Mawby, Professor Joanna Shapland, Professor Nick Tilley, Professor van Dijk and many others. The editors of the journal are Dr Martin Gill and Adrian Beck from the internationally renowned Centre for the Study of Public Order at Leicester University, UK.

The journal welcomes articles, research papers, commentaries, etc. for publication. Contributors are welcome to discuss ideas with the editor (0116 252 5703/9).